Contemporary Problems in Geography

The general editor of *Contemporary Problems in Geography* is Dr. William Birch, who is Director of the Bristol Polytechnic. He was formerly on the staff of the University of Bristol and the Graduate School of Geography at Clark University in the U.S.A. and he has been Chairman of the Department of Geography in the University of Toronto and Professor of Geography at the University of Leeds. He was President of the Institute of British Geographers for 1976.

Alan Wilson is Professor of Regional and Urban Geography at the University of Leeds. After reading mathematics at Cambridge he has served as Scientific Officer at the National Institute of Research in Nuclear Science, Research Officer at the Institute of Economics and Statistics, University of Oxford, Mathematical Adviser at the Ministry of Transport, and Assistant Director of the Centre for Environmental Studies. His publications include *Entropy in Urban and Regional Modelling*, *Papers in Urban and Regional Analysis*, and *Urban and Regional Models in Geography and Planning*.

Michael Kirkby is Professor of Physical Geography at the University of Leeds. He has done research at Cambridge and at the Johns Hopkins University and previously taught at the University of Bristol. In 1972 he published *Hillslope Form and Process* with M. A. Carson.

Environmental Change

ANDREW GOUDIE

CLARENDON PRESS · OXFORD
1977

Oxford University Press, Walton Street, Oxford OX2 6DP

OXFORD LONDON GLASGOW NEW YORK
TORONTO MELBOURNE WELLINGTON CAPE TOWN
IBADAN NAIROBI DAR ES SALAAM LUSAKA ADDIS ABABA
KUALA LUMPUR SINGAPORE JAKARTA HONG KONG TOKYO
DELHI BOMBAY CALCUTTA MADRAS KARACHI

© Andrew Goudie 1977

British Library Cataloguing in Publication Data

Goudie, Andrew
   Environmental change.—(Contemporary problems
in geography).
   1. Physical geography   2. Historical geology
   3. Climatic changes
   I. Title   II. Series
   574.5   GB53   77–30249

   ISBN 0–19–874073–5
   ISBN 0–19–874074–3 Pbk

Filmset in 'Monophoto' Times 10 on 12 pt and
printed in Great Britain by
Richard Clay (The Chaucer Press), Ltd,
Bungay, Suffolk

Editorial Preface

The progress of research in physical geography has been accompanied, over the last twenty years, by a progressive shift in teaching content at all levels. The changes have filtered down from final year to first year courses in higher education, and increasingly to sixth form work. Suitable textbooks have been rather slow to follow, and it is only in the last five years that the trickle of advanced-level texts has grown to a relative flood. First and second year undergraduate courses are still poorly served, even though the need for course texts is much more important at this level.

Contemporary Problems in Geography is a series which is intended to fill this gap for geography as a whole; and this is one of three volumes which are being published this year on systems of interest to physical geographers. Andrew Goudie's book *Environmental Universe* is an introduction to the framework of recent earth-history which is a crucial context for all studies of current process or landform. Ian Statham's book *Earth Surface Sediment Transport* examines the processes which move sediment on hillslopes and in rivers, and so form the landscape. Steve Trudgil has written *Soil and Vegetation Systems*, a book on the processes by which the plants and soil interact to provide the conditions for plant growth, and determine the course of soil evolution.

All three books are intended for first and second year undergraduates who have done little or no specialized physical geography before, and who need a broad understanding which will fit in with other aspects of geography or environmental science.

March 77

Preface

The study of the environmental changes of the last three million years, which forms the subject of this book, can be approached from a wide variety of viewpoints. The viewpoint adopted here is that of the geographer—the aim being to show how the physical environment and landscape on the Earth have changed during the time that man has been living on the Earth, and to suggest, by example, some ways in which the great environmental changes may have influenced his development.

These environmental changes include not only those of climate, but also changes of sea-level, of vegetation associations, desert limits, lake levels, river discharges, hurricane frequencies, sea-ice cover, numbers of mammals, and many others. Particular attention is paid to the degree and frequency of change, for it is often insufficiently appreciated how frequent and substantial the changes have been, even within historical time. To understand the nature and origin of present-day soils, landforms, and floral and faunal distributions, it is essential to be aware of their history and evolution. Many features of the environment and of the landscape are not necessarily in equilibrium with present processes, and thus it is frequently inappropriate to examine them purely in terms of currently functioning systems.

The changes which man himself has made to his environment and landscape, however, only form an incidental part of this book. This is not because the influence of man is considered to be unimportant—the reverse is the case—but because it could in itself form the basis of a lengthy volume. However, particularly in the last century or so, man has become an increasingly potent agent of environmental change, and to put 'natural' environmental changes into context, some reference has had to be made to these developments.

Another feature of this volume is that, inevitably, the intensity of treatment of the various phases of the last three million years increases as one moves towards the present. This reflects two underlying facts: our knowledge becomes less uncertain, and our chronology more exact, as the present is approached; and the relationships of environmental changes to human affairs is more evident, not least because of the exponential growth of human population. Nevertheless, this book is not meant to be crudely deterministic. All it seeks to do is to show that changes have taken place in man's environment to a remarkable degree, and to indicate some of the ways in which such changes may have been related to the development of man and the landscape.

During the preparation of this work I have been conscious of the debt which I owe to numerous teachers from my undergraduate days. At Cambridge the degree of dedication to the Quaternary is considerable, and this once led an Oxford visitor to make the quip that 'If the Ice Ages had not existed Cambridge would have had to invent them so as to have something to lecture on.' Soil stripes in the Breckland, the ice wedges of Fen margin gravel pits, the fossil frost mounds of the Lark valley, the organic deposits of the West Runton section, and the inland dunes at Lakenheath Warren were part of one's undergraduate diet. The post-graduate diet was scarcely more varied, but the flavours were hotter, for I was offered the opportunity to study environmental changes in the Kalahari sandveld, the Namib desert, the lakes region of the southern Ethiopian Rift Valley, and the fringes of the Thar. At an even earlier stage I owe much to P. G. Foster, E. S. Hoare, M. A. Girling, and C. Kenyon. Without their stimulus, flair, and dedication I would not have become a geographer.

Amongst those I would like to thank for giving me field experience are Dick and Jean Grove, Bridget and Raymond Allchin, and K. T. M. Hegde. I should also like to thank David Stoddart for encouraging me to write this book, Councillor John Patten for tolerating my early efforts to produce this volume in a Woodstock cottage, the Librarians of the School of Geography and the Radcliffe Science Library for their resourcefulness and assistance, and Peter Masters and Chris Jackson for helping in a multitude of tasks. The Figures were kindly drawn by Mrs. Ursula Miles and Miss Margaret Loveless, and I received some helpful comments from C. G. Smith, Fellow of Keble College. Professor Gordon Manley, and Miss Alayne Street. Some of the material in this book was tried out on a remarkable trio of Hertford undergraduates, Mary Francis, Ken Pye and John Johnson, to whom go my best wishes.

Oxford.   A. S. GOUDIE

Contents

x   Contents

1 Introduction

We have in the history of the Quaternary Period a
region of research, full of hope and of the most roman-
tic interest, promising not only to reveal the events
which accompanied and influenced the evolution of man
but to afford an outpost from which we can look back
into the ages which preceded his advent upon the Earth.

W. B. WRIGHT (1937, p. 464)

Environmental change during the time of man

Man, sometimes called the 'tool-making animal', has only been an inhabitant of the Earth for a fraction of its history. Latest estimates suggest that
whereas the Earth as a planet is around 5000 million years old, man has
only been evident for the last 2 to 3 million years. Over many parts of the
world man's arrival is even later than this. For example the earliest dated
remains of human beings found in Australia are only around 30 000 years
old, whilst those from the New World are seldom older than 15 000 years
(though some earlier dates have now been claimed). Inhabitation of New
Zealand, Madagascar, and Oceania came even later. The oldest records of
human activity, crude stone tools which consist of a pebble with one end
chipped into a rough cutting edge, have been found in conjunction with
bone remains from various parts of the African continent (Leakey &
Goodall, 1969). At Lake Rudolf in northern Kenya, and the Omo Valley in
southern Ethiopia, a tool-bearing bed of volcanic material called tuff has
been dated by the new isotopic techniques at about 2·6 million years old,
while another bed at the Olduvai Gorge in Tanzania has been dated by
similar means at 1·75 million years.

This book is concerned with the history of the time during which man has
lived on the Earth, a period which the geologist calls the Quaternary. The
Russians, mindful that this period was unique because of man's presence,
have sometimes called it the *Anthropogene*. While this period may have been
very short in relation to the total age of the Earth (Figure 1.1) the environmental changes which have occurred during the time that man the tool-
maker has been an inhabitant of the Earth have not been slight. Changes,
whether in climate, sea-levels, vegetation belts, animal populations, or soils
and landforms, have been both many and massive. Some changes are still
taking place, at various speeds, and at various scales. Even if one rejects
some of the more extreme proposals as to the effects that such changes have
had on man and his history, the changes have been of such an order as to

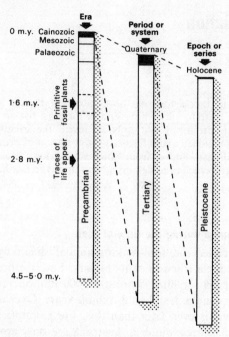

FIG. 1.1. The Quaternary and its subdivisions in relation to the geological time-scale (from Vita-Finzi, 1973, Fig. 1 and other sources).

markedly influence both man and landscape. In the last 20 000 years alone, and man, as already noted, has been an inhabitant of parts of the Earth for at least a hundred times longer than that, the area of the earth covered by glaciers has been reduced to one-third of what it was in a glacial maximum, the waters thereby released have raised ocean levels by over one hundred metres, the land, unburdened from the weight of overlying ice, has locally risen by as much as several hundred metres, vegetation belts have swung through the equivalent of tens of degrees of latitude, inland lakes have expanded and contracted, desert sand fields have advanced and retreated, and many of the finest mammals have perished in the catastrophe called 'Pleistocene overkill'.

Even at the present time, smaller fluctuations of climate are leading to changes in fish distributions in northern waters, marked fluctuations in valley glaciers, extensive flooding of African lakes, and difficulties for major agricultural schemes in Central Asia.

Thus environmental changes are a basic concern in a consideration of the relationships between man, the environment, and landscape. At the same time they have become a major focus for the attention of participants in many major disciplines, including economic historians interested in price fluctuations in medieval times, archaeologists concerned with the rise and fall of cultures and civilizations, geomorphologists conscious of the imprint of past processes on landforms, biologists, botanists, and zoologists concerned with the evolution, speciation, and distribution of organisms, geologists involved in the sedimentary and stratigraphic record, and oceanographers occupied with changing water masses, currents, and salinities in the world's oceans.

The terminology employed to describe some of the events with which we are concerned, perhaps needs to be stated at this point. The major geologic-time divisions currently employed are as follows: the latest *era* is called the Cainozoic, and this has been divided up into two *periods*; the Tertiary and the Quaternary. These *periods* are in turn divided up into various *series* or *epochs*, with the Pliocene being the final series of the Tertiary. The Pleistocene and Holocene series together constitute the Quaternary. It is exceedingly difficult to draw boundaries between some of these units. Notably this is the case between the Tertiary and the Quaternary. As a consequence some authorities currently consider the Tertiary and the Quaternary as constituting the Late Cainozoic. Likewise, because the boundary between the Pleistocene and the Holocene is no different in character to the boundaries between the various preceding glacials and interglacials, some authorities would consider the Holocene as the most recent stage of the Pleistocene rather than as a separate series. Other arguments against retaining the term Holocene as a series name are that the Pleistocene is still in progress, and that the Holocene has been too brief to merit the status of a series or epoch. These problems have been discussed in a series of works (see, for example, Flint, 1971, p. 379; West, 1968, p. 224; and Vita-Finzi, 1973, p. 35). In this book the terms Pleistocene, Quaternary, and Holocene are used in the conventional way, largely because of their familiarity.

The scale of environmental change

Changes in environment have occurred over a wide range of time spans. Table 1.1 is a summary of the main orders of climatic change which can be identified, from the minor fluctuations within the period of instrumental record, which occur within a time scale of around a decade, to the major geological periods, with durations of many millions of years. Major phases of ice-age activity, for example, appear to have been separated by around 250 million years.[1] As an instance, there is unequivocal evidence that between 350 and 250 million years ago there were glaciers in modern equatorial regions, whereas, between 200 and 100 million years ago, the Earth experienced a period of more even climatic conditions, generally warmer than today. It is likely that changes on this massive time-scale have been caused in part by changes in the positions and configurations of the continental

[1] Brooks (1926) identified a 250 million year cycle of major glaciations which he saw as 'episodes disturbing the normal climate for a brief time, as at long intervals a passing cyclone disturbs the peaceful life of a tropical island'. Subsequent work suggests that this model may be too simple.

Table 1.1
Orders of climatic variation

	Time-scale unit	
(1) Minor fluctuations within the instrumental record	10 years	Minor fluctuations which give the impression of operating over intervals of the order of 25–100 years, with somewhat irregular length and amplitude
(2) Post-Glacial and historic	10^2 years	Variations over intervals of the order of 250–1000 years, e.g. the sub-Atlantic recession and others affecting vegetation in Europe and N. America
(3) Glacial	10^4 years	The phases within an ice age, e.g. the duration of the Würm was of the order of 50×10^3 years
(4) Minor geological	10^6 years	Duration of ice ages as a whole; periods of evolution of species
(5) Major geological	10^8 years	e.g. ice ages at intervals of 2.5×10^8 years

Principal bases of evidence

(1)   Instrumental; behaviour of glaciers; records of river-flow and lake levels; non-instrumental diaries; crop yields; tree-rings (also for dating).

(2)   Earlier records of extremes; fossil tree-rings; archaeological finds; lake levels; varves and lake sediments; oceanic core-samples; pollen-analysis; radiocarbon dating.

(3)   Fauna and flora characteristic of interglacial deposits; pollen-analysis; variation in height of snowline and extent of frozen ground; oceanic core-samples (dating through latter).

(4, 5) Geological evidence; character of deposits, fossil fauna and flora; dating largely through radioactivity of rocks.

(After Manley, 1953)

masses brought about by sea-floor spreading and plate tectonics. The ice ages themselves, in contrast, may, as in the case of the Pleistocene ice age, have lasted around 2·5 million years, while the various phases of glacial advance within each ice age may have been only 50 000 or so years long. It is not possible to envisage such short-term changes as having been caused by continental drift, and some other mechanisms need to be considered (see Chapter 7). It is these shorter-term changes which form the most important events on the backcloth of human history.

The evolution of ideas
Appreciation of the fact that the world had such a history of environmental change emerged at much the same time that it was appreciated that the world had a history of some length beyond the constricting 6000 years of Archbishop Ussher's biblical time span. This concept, which saw the Earth as having been created in 4004 B.C., was relatively little contested until the end of the eighteenth century, and it was accompanied by the belief that much of the denudation and deposition evident on the face of the Earth could be explained through the agency of Noah's Flood and other catastrophic events. Gradually, however, these ideas were shown to be erroneous through the evidence collected by geologists and natural historians like

Guettard, Count Buffon, and George Poulett Scrope. In particular, James Hutton and his friend John Playfair, the Edinburgh scientists, are often regarded as the men who most effectively propagated the new ideas, for in the geological record they saw 'No vestige of a beginning, no prospect of an end'. They realized that the complexity of the sedimentary record could be explained through the operation of processes akin to those of the present day extending over a long time span.

The idea that climate and other aspects of the environment had fluctuated or changed during this enlarged span of time resulted initially from the discovery that Norwegian and Alpine glaciers had formerly extended further than the bounds of their current limits. Some suggestions as to this had originally been made by scientists at the end of the eighteenth century. In 1787 de Saussure recognized erratic boulders of palpably Alpine rocks on the slopes of the Jura Ranges, and Hutton reasoned that such far-travelled boulders must have been glacier-borne to their anomalous positions. Playfair extended these ideas in 1802, but it was in the 1820s that the *Glacial Theory*, as it came to be known, really became widely postulated. Venetz, a Swiss engineer, proposed the former expansion of the Swiss glaciers in 1821, and his ideas were supported and strengthened by Charpentier in 1834. The poet Goethe expressed the idea of 'an epoch of great cold' in 1830. However, the ideas of both Venetz and Charpentier were extended and widely publicized by their fellow countryman, Louis Agassiz, who was one of the originators of the term *Eiszeit* or *Ice Age*. In Norway Esmark put forward similar ideas in 1824, and in 1832 Bernhardi went so far as to suggest that the great German Plain had once been affected by glacier ice advancing from the North Polar region.

In spite of this convergence of opinion from numerous sources, the ideas of these original minds were not easily accepted or assimilated into prevailing dogma, and for many years it was still believed that glacial till, called drift, and isolated boulders, called erratics, were the result of marine submergence, much of the debris, it was thought, having been carried on floating icebergs. Sir Charles Lyell noted debris-laden icebergs on a sea-crossing to America, and found that such a source of the drift was more in line with his belief in the power of current processes—the uniformitarian belief—than a direct glacial origin. For years even glacial or subglacial depositional features like eskers [1] were thought to be of a marine origin, and were classified into fringe eskers, bar or barrier eskers, and shoal eskers. Moreover, in Britain, some drift deposits contained marine shells, a fact that gave prima facie support to marine ideas.

One of the first converts to Agassiz's *Eiszeit* concept was Oxford's Dean Buckland, who was won over when Agassiz visited Britain in 1840. Certain

---

[1] Eskers are long, narrow ridges, chiefly of gravel and sand, which once formed the beds of streams flowing in the ice of a glacier, usually at the bottom, and which were left when the ice melted. They are very widespread in lowland Ireland.

other great geologists, like Sir Roderick Impey Murchison, were slow to accept the new ideas on extensive glaciation, but he, too, in the 1860s was in due course persuaded that his native Scotland and other parts of Europe had been glaciated, and that much of the drift or till resulted from their action. Some people remained unconvinced. In 1892, for instance, H. H. Howorth produced his massive neo-catastrophist *The Glacial Nightmare and the Flood—a second appeal to common sense from the extravagance of some recent geology*, and tried to return to a fundamentalist-catastrophist interpretation of the evidence.

Agassiz himself went on to a professorship at Harvard later in the 1840s, and thereby helped to spread glacial ideas in North America, though some American workers, including Conrad and Hitchcock, appear to have embraced the concept before his arrival.

The study of environmental change progressed somewhat further in the 1860s when Sir Archibald Geikie, A. C. Ramsay, and T. F. Jamieson showed that the spread of the glaciers had not been a single event, and that there had been various stages of glaciation, represented by moraines, which had been separated by warmer phases, called *interglacials*. These interglacials can be defined as a particular type of non-glacial climatic condition with a climatic optimum at least as warm as that experienced during the Holocene. Another type of interruption to full glacial conditions was the *interstadial*, a period which was either too cold or too protracted to allow the development of temperate deciduous forest of the full interglacial type. The term *stadial* refers to an ice advance.

The concept of multiple glacials, interglacials, stadials and interstadials, during the Ice Age was further confirmed in the first years of the twentieth century by Penck and Brückner (1909), working in the Alps (see pp. 26–28). They developed much of the terminology and interpretation of sequences used to this day.

In areas outside those which were subjected to Pleistocene glaciation other types of fundamental environmental change were recognized as having taken place, though Louis Agassiz, after a journey to Brazil, postulated that even equatorial regions had undergone glaciation. He believed that the Amazon Basin had been overwhelmed, but seems, like certain others at the time, to have allowed his enthusiasm to convince him that what was in reality deeply weathered soil, and boulders produced by intense chemical activity under tropical conditions, was glacial boulder clay. A truer appreciation of the effects of climatic change in non-glaciated regions of the tropics and subtropics came from Jamieson, Lartet, Israel Russell, and Grove Karl Gilbert, all of whom studied the fluctuations of Pleistocene lakes in semi-arid areas, as represented by their old strandlines, deltas, and algal limestones. Their work in effect established the supposed general relationship between high-latitude *glacials* and mid- and low-latitude *pluvials*, and Russell was able to tie in the moraines of the former Sierra

Nevada glaciers with old strandlines around Lake Mono in California. Gilbert demonstrated that Lake Bonneville, in much the same way as the glaciers, had fluctuated several times, with alternations of high water and phases of desiccation. Other geologists investigating the west of the United States also believed that a diminution of rainfall might account for some of the anomalous drainage features they encountered on their travels, and the term *pluvial* was itself coined by Alfred Tylor in 1868. It means basically a period of comparatively abundant moisture in regions beyond the limits of the ice caps. Whether or not such pluvial periods were in phase or not in phase with the high-latitude glacial periods is a controversial problem referred to on page 83.

Comparable in importance to the climatic changes of the Pleistocene, and associated in large measure with them, were the great changes in the relative levels of land and sea. Once again the name of Playfair is an important one, for in his *Illustrations of the Huttonian theory of the Earth* (1802) he made detailed descriptions of the emerging shorelines of Fennoscandia, and assessed correctly that the cause in that particular case was crustal uplift, though this was then supposed to result from progressive cooling of the crust being accompanied by shrinkage. However, it was Jamieson working in Scotland in the 1860s who first proposed that the weight of glacial ice sheets might cause subsidence of the crust beneath them, and that the release of this weight by melting in post-glacial times would lead to uplift. Gilbert extended this idea, and attributed the observed warping of the pluvial Lake Bonneville shorelines to removal of the weight of water on desiccation, whilst another North American worker, Dutton, was perhaps the most successful person in indicating the importance of such processes, to which the name *isostasy* has been given.

Many years earlier, however, in the 1830s and 1840s, Lyell and Maclaren had argued that more extensive glaciers and ice sheets would have stored up considerable bodies of water, and that, as a consequence, the level of the sea must have been many metres lower than at present. This was the *eustatic* theory of 'swinging sea-levels'. The causes and effects of sea-level changes, produced by this and other mechanisms, are discussed in the penultimate chapter of this book.

Traditional techniques

Although some new techniques have now become highly important, they have supplemented rather than replaced some of the more traditional methods for determining the chronology and environments of the Quaternary. Of these older techniques, the study of varves, tree-rings, pollen, plant macro-fossils, and faunal remains, have been especially significant.

Varves are regular alternations in a sediment of layers of different composition or texture, forming pairs or couplets, attributable in general to an annual seasonal rhythm. Such layers were produced by the great summer inputs of meltwater from glaciers into lake basins. The coarser material was deposited in the summer, but the fine suspended clays in the meltwater did not settle out until the autumn or following winter. It is this which produces the banding. Some years would tend to produce particularly thick bands, and varve chronology depends on the correlation of such distinctive bands in different localities, and on the assumption that the couplets are annual. This technique was pioneered with great effect by de Geer in Sweden, and results have sometimes been substantiated by radio-carbon dating (Tauber, 1970). Where material datable by C14 is scarce or unavailable this technique still has a degree of utility. In a sense it utilizes the same principle as tree-ring analysis (dendrochronology), whereby attempts are made to correlate the annual growth rings of trees from area to area. In favourable situations tree-rings may be related in their growth to precipitation levels and can thus be used for climatic reconstruction (see Fig. 1.2). The time range covered can extend back some three or four thousand years where trees of such ages survive, as is the case with the Bristlecone Pine of the south west U.S.A. The nature of the rings also affords the possibility of dating structures in which

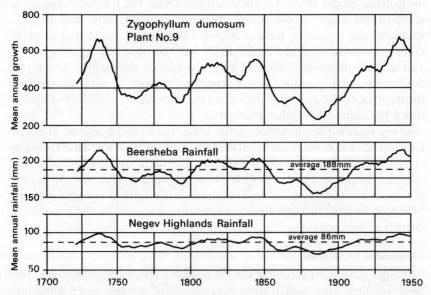

FIG. 1.2. Tree-rings and derived annual rainfall in the central Negev, Israel, 1720–1950.
Annual growth of the woody shrub, *Zygophyllum dumosum*, in the central Negev, Israel, has been shown to give reliable estimates of the average rainfall of Beersheba and thus also for the Negev highlands. Existing 45-year rainfall records of Beersheba were used to estimate the average annual rainfall for the central Negev highlands. The 25 year rainfall records were then derived, as shown in the above curves (21-year moving weighted means for the period 1720–1950) (from Shanan, Evenari, and Tadmor, 1967).

tree remains are included and has provided an independent check on the validity of C14 dating.

Pollen-analysis, or palynology, is one type of micro-fossil analysis. It makes use of the fact that some sediments contain pollen grains and spores which mostly come from the air by fallout (pollen-rain), and they are thus derived from the regional and local vegetation. Vegetational changes, which may be caused by climatic, edaphic, or biotic factors, may be recorded by the preservation of pollen in a section. Pollen grains may be counted and recorded by dispersing sediments with appropriate agents and then looking at them under a high-power binocular microscope. Results of such analysis give a picture of the vegetation at a given point in time, and also allow the sequence of vegetational change to be examined over a period.

An equally laborious technique, but one which has produced good results, is that which utilizes non-marine molluscs, remains of which are found very commonly in Pleistocene deposits. Molluscan assemblages have been found to be indicative of particular types of climate. Cold faunas show a dominance of a few species in great numbers, temperate faunas a larger number of species, many of which occur frequently.

Similarly, especially at the University of Birmingham, techniques have been developed to use remains of beetles (Coope *et al.*, 1971). Their wing cases are found in suitable sediments, and as distributions of living species are known quite well, especially in Scandinavia, it has proved possible to interpret palaeo-environments on the basis of insect faunas. Results have tended to correlate well with results obtained by pollen-analysis, and by the study of non-marine mollusca, though certain discrepancies have been noted. At Lea Marston, Warwickshire, for example, a deposit dated to about 9500 B.P. (before present) shows a relatively warm beetle fauna but a cool vegetation assemblage dominated by *Betula*, *Salix*, and some *Pinus*. This appears to be because the very mobile insect fauna was able to react to the very rapid climatic amelioration around that time in advance of the more slowly migrating trees (Osborne, 1974).

New chronological techniques

In the last couple of decades, however, the study of environmental change has been greatly transformed by the development of certain new techniques, particularly for dating, and for temperature assessment. These have enabled more accurate absolute dating of events over an extended time-scale, facilitating both temporal and spatial correlations which, until then, had been extremely hazardous. These techniques, combined with the spread of detailed scientific exploration in areas hitherto neglected, notably parts of South America, the Kalahari, Ethiopia, India, and the Polar regions, have led to many changes in our view of the world's history since its occupation by man.

Of especial interest have been the various radiometric (isotopic) dating

l Sotopic Dating

techniques especially radiocarbon, uranium series, and potassium-argon. Some of the features of these methods are shown in Table 1.2.

These three isotopic dating techniques all depend on the measurement of amounts of elements which through time are either formed by, or are subject to, radioactive decay. The rate of decay being known for a particular element, the time-interval may be assessed between the present, and the time when the particular parent material was fixed and its decay began. Thus, for

Table 1.2
Some isotopic methods of dating Quaternary deposits

Name	Isotope	Half-life (years)	Range (years)	Materials
Radiocarbon	C14	5730 ± 40	0–50 000	Peat, wood, shell, charcoal, organic muds, algae, tufa, soil carbonates
Uranium Series	U^{234}	250 000	50 000–1 000 000	Marine carbonate, coral, molluscs
	Th^{230}	75 000	0–400 000	Deep-sea cores, coral, and molluscs
	Pa^{231}	32 000	5000–120 000	Coral and molluscs
Potassium–argon	K^{40}	1.3×10^9	greater than 20 000	Volcanic rocks, granites, etc.

example, a growing organism incorporates radiocarbon, and on its death the radiocarbon is trapped and then begins to decay. As half the radio-activity will be lost after an interval calculated to be about 5730 years, by measuring the radioactivity of fossil material containing carbon, the date at which death took place can be determined. Radiocarbon or C14 dating, formerly used mainly for organic carbonate, in the form of peat and wood, is now being extended to a wider range of Late Pleistocene materials, especially soil carbonates and mollusca. This technique has evolved steadily since its first application in 1949, and it provides, together with archaeological evidence, the chronology for approximately the last 60 000 years though practical problems become severe beyond about 40 000 years B.P.

Useful though it may be, radiocarbon dating still has many problems which need to be considered in assessing the reliability of the very large number of dates which are now available. In particular, contamination of samples may take place. Humic acids, organic decay products and fresh calcium carbonate may be carried downwards to contaminate underlying sediments. In the case of inorganic carbonates 'young' carbonate may be precipitated in, or replace, the carbonate which one is interested in dating, and removal of the contaminant from pore spaces and fissures is almost impossible. Additional to problems such as these, are miscellaneous other problems, including the fact that different laboratories may use different half-lives, and the discovery that fluctuations in cosmic radiation with time may produce slight differences in the C14 equilibrium of the atmosphere, biosphere, and hydrosphere.

Since the early 1960s potassium-argon (K/Ar) dating has been applied to Pleistocene and Pliocene chronology and, as described on p. 19, has greatly changed our views on the length of the Pleistocene, and on the time when glaciation was initiated. Whilst radiocarbon dating utilizes organic and inorganic carbonates, potassium-argon dating, which can cover a theoretically unlimited timespan, utilizes unaltered, potassium-rich minerals of volcanic origin in basalts, obsidians, and the like. It is, however, only usable in practice for materials older than about 50 000 years.

Also in the 1960s the Thorium-Uranium and other Uranium Series dating methods have been applied to such materials as molluscs and coral. Although still subject to certain deficiencies, particularly for molluscs, these techniques are extremely valuable when applied to coral in bridging the gap between radiocarbon and potassium-argon techniques. These methods have been used for materials up to about 200 000 years in age with some success, and Uranium Series dates obtained for coral terraces have caused a change in ideas on the fluctuations of sea-level before the Last Glaciation (see p. 180).

In addition to the isotopic techniques, great use has recently been made of a palaeomagnetic calendar of magnetic events. Currently the Earth has what is termed a 'normal' magnetic field so that at the north magnetic pole a compass dips vertically towards the Earth's surface. However, the magnetic field, for reasons not fully understood, can switch to become 'reverse'. As some rocks and sediments may preserve the characteristic signal of the magnetic field during the time the unit was deposited, it has been possible to produce a calendar of magnetic events marked by switches from 'normal' to 'reverse'. As many of these switches have now been dated by independent means, in a conformable sequence of sediments these magnetic switches enable a particular section to be dated against a master system (Glass *et al.*, 1967). Thus sediments from deep-sea cores can be given an age-scale of considerable length.

A two-level system of names has been introduced to describe the observed sequence of polarity reversals. At the lower end are *polarity events*—short intervals of normal or reversed polarity lasting in the order of 150 000 years or less. At the higher level are the *polarity epochs*—longer intervals during which the magnetic field was predominantly of one polarity, and which may contain one or more events (Cox *et al.*, 1968). The dates of these epochs and events are shown in Fig. 1.3.

Volcanic eruptions can also provide important stratigraphic markers for the Quaternary. Different ash falls may be recognizable on the basis of

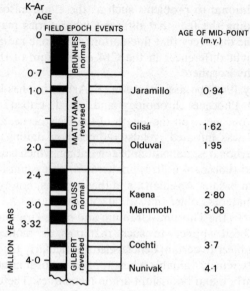

Fig. 1.3. Time-scale for geomagnetic reversals. Times when the field was normal are indicated by shading (after Cox, Doell and Dalrymple, 1968, Fig. 1).

petrology and chemical composition. The falls of ash are placed in chrono-stratigraphic position by C14 dating of associated sediments, or by K/Ar dating of the source volcanic unit. Once the age has been established, an ash can be used as a marker horizon in otherwise undated sections. This technique is termed tephrochronology.

One other dating technique needs mention: lichenometry. This has become increasingly important in the last decade, and is especially useful for dating glacial events over the last 5000 or so years. It is believed that most glacial deposits are largely free of lichens when they are formed, but that once they become stable, lichens colonize their surfaces. The lichens become progressively larger through time. Thus by measurement of the largest lichen thallus of one or more common species, such as *Rhizocarpon geographicum*, an indication of the date when the deposit became stable can be attained.

Developments in ocean core, lake, and cave stratigraphy

Of comparable importance to the new dating techniques has been the development of deep-sea coring procedures, for although by no means completely stable (burrowing organisms and currents creating problems) the sea floor does offer a more continuous and lengthy stratigraphic record than most terrestrial sections. It is from the evidence of the deep-sea cores that a series of cold and warm episodes can be dated, identified, and perhaps related to, glacials and interglacials. The cores, normally obtained by piston corers, have also helped to establish the age of the Plio-Pleistocene boundary—formerly a matter of great dispute. The deep-sea cores can be used and interpreted in a variety of ways. Core materials can be dated by radiometric means, palaeomagnetic epochs (see p. 11), whether normal or reversed, can be identified in the sediment layers, microfossils (especially Foraminifera and Radiolaria) can be examined, and the lithological characteristics of the sediments within the cores can be determined with a view to finding out about changes in terrigenous sediment sources.

One of the most productive ways of examining the cores has been the study of changes in the frequency of particular 'sensitive species' Foraminifera. These are thought to reflect changes in the temperature of ocean waters (Kennett, 1970). *Globorotalia menardii* tests, for example, may be counted, and the ratio of their number to the total population of Foraminifera tests can be worked out. The ratio can range from as high as 10 or 12, to as low as nearly zero. A high ratio appears to be associated with the warm water of interglacial conditions, while a low ratio appears to be associated with the colder water of glacial stages. Thus, by taking samples along the length of a core, the alternations of warmth and cold can be established. Similarly, another of the Foraminifera, *Globorotalia truncatulinoides*, can be used to the same end. In any portion of a core-sample some of its tests will show a left-hand direction of coiling, and others will show a right-hand direction of coiling. It has been found by some workers that right-coiling tests are associated with warmer conditions, and left-coiling tests with cooler. Thus ratios of left to right coiling may enable an assessment to be made of palaeoclimates. Some workers have attempted to use more sophisticated methods of utilizing Foraminifera, and instead of approaching the problem through the study of 'sensitive species', they have tried to establish climatic sequences based on 'total fauna' (see Shackleton, 1975 for a discussion).

Another way in which the Foraminifera can be utilized is by the measurement of the O^{18}/O^{16} ratios in the calcitic tests. This oxygen isotope method was developed in the 1950s, by Emiliani and others (Emiliani, 1961). He supposed that the O^{18}/O^{16} ratio depended substantially on the temperature of the water in which the Foraminifera lived. While there is now some controversy as to the value of this technique in giving quantitative data on palaeotemperature changes (Shackleton, 1967), the method does appear to give a fairly clear picture of the periodicity of the major glacial and interglacial episodes, and it has helped to show that the Pleistocene was characterized by more glacial cycles than had been suspected on the basis of evidence from the terrestrial record. However, the ice sheets have played an important role in determining the oxygen-isotope record. During the glacial episodes immense ice sheets of isotopically light ice accumulated in northern America and Europe. When this occurred, the oceans diminished in volume, became slightly more saline, and became isotopically more positive (i.e. enriched in O^{18}) (Shackleton, 1975). Also obtainable from the deep-sea core

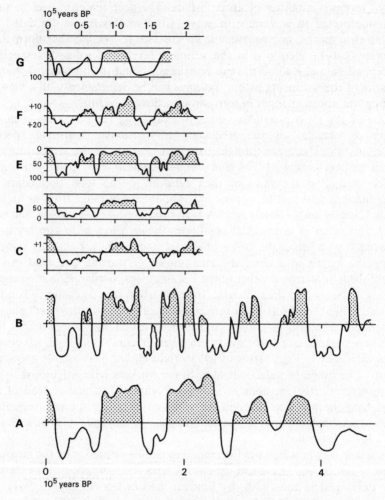

Fig. 1.4. Fluctuations in climate over the last 400 000 years as revealed by miscellaneous indicators from ocean cores. Shaded areas = warmth, non-shaded = cold.

A. Generalized curve for the Equatorial Atlantic based on ratios of cold to warm Foraminifera (after Bowles, 1975, Fig. 4).

B. Generalized curve for the Polar Front in the NE. Atlantic based on ratios of cold to warm Foraminifera (after Bowles, 1975, Fig. 4).

C. Oxygen-isotope curve for a Caribbean core at 15 °N (P6304–9) (after McIntyre *et at.*, 1972, Fig. 6).

D. Curve of percentage coccolith carbonate for an Atlantic core (V-23-83) at 51 °N (after McIntyre *et al.*, 1972, Fig. 6).

E. Curve of percentage polar fauna (Foraminifera) for Atlantic core at 51 °N (V-23-83) (after McIntyre *et al.*, 1972, Fig. 6).

F. Curve based on oxygen-isotope variation at core 280A (40 °N) in the Atlantic (after McIntyre *et al.*, 1972, Fig. 6).

G. Curve based on the weight (per cent) of glacial detritus in the sand fraction of core RE5-36 at 50 °N in the Atlantic (after McIntyre *et al.*, 1972, Fig. 6).

evidence is the extent to which iceberg-rafted debris is present. In middle-latitudes this is an indirect indicator of cold climate, though in high latitudes ice-rafted debris maxima may be associated with interglacial periods (Keany *et al.*, 1976). This technique was extensively applied in the 1960s, notably to the North Pacific (Kent *et al.*, 1971), the Southern Ocean (Opdyke *et al.*, 1966), and the Arctic (Hermann, 1970).

When one compares the results obtained from these different approaches to gaining palaeoclimatic information from deep-sea cores, one often finds a remarkable degree of similarity in the pattern of the curves, especially for the upper portions of cores. This is illustrated by Fig. 1.4 which shows curves derived from oxygen isotope studies, from the proportion of glacial material, from the amount of carbonate, and from the frequencies of polar Foraminifera.

Comparable to such sedimentological evidence is that provided by the presence of aeolian debris on the ocean floors (Parmenter and Folger, 1974). This, together with the presence of large quantities of unweathered minerals, including feldspars, has been used to assess whether tropical climates were dominantly arid and semi-arid, or whether they were dominantly humid (see p. 83) during particular phases. During arid phases rivers would tend to carry unweathered feldspars, while under more humid conditions the quantity of feldspars relative to the more stable quartz would be less. Similarly, in oceanic sediments off west Africa, opal phytoliths, and freshwater diatoms are common in sediments deposited when waters were warm, but less common in sediments deposited when waters were cold (Parmenter and Folger, 1974).

Cores have also been put down into the floors of lakes both in temperate and in tropical areas. These can indicate changes in the nature of sediments deposited over a long time-sequence. In some tropical lakes, for example, evaporite layers may be identified, and regarded as being the product of dry conditions (see, for instance, Kendall, 1969). Alternatively the core-samples can be subjected to sophisticated chemical analysis (Degens and Hecky, 1974). In Lake Kivu, East Africa, for instance, Fe–Ni sulphide contents were thought to indicate strongly reducing conditions, and thus high water stands, while high Al and Mn contents were thought to indicate low stands. As some lake cores may be several hundred metres thick, these types of investigations can extend back a considerable time.

Some success has been attained through the isotopic study of carbonate-rich cave sediments. Their history, and the temperature conditions associated with them, have been studied, for example, in France (Duplessy, 1970), and in New Zealand (Hendy and Wilson, 1968) using Th^{230}/U^{234} techniques.

Much of the data that have been obtained by these methods have recently been used to try and reconstruct the nature of world environmental conditions at certain selected points in the Pleistocene (Climap Project Members,

1976), and attempts have been made to simulate conditions on computers with a global atmospheric model (Gates, 1976). Such attempts at reconstruction and simulation on such a broad scale are in their infancy, but may do much to increase our understanding of the past, and, it is hoped, enable some prediction of the future.

Ice cores

In the last few years the lengthy record provided by the deep-sea record has been supplemented by deep cores obtained from the ice caps, at Byrd Station in Antarctica, at Devon Island in Arctic Canada, and at Camp Century and Crête in Greenland. The Camp Century Core (Epstein *et al.*, 1970), put down in north west Greenland, attained a length of no less than 1390 m. It represents a seemingly continuous sequence of annual layers of former snow.

The core was sampled at regular intervals by Dansgaard and his co-workers (1969) for its O^{18}/O^{16} ratio. This ratio depends mainly on the temperature of condensation at the time of ice formation. Thus a plot of the O^{18}/O^{16} ratio down the length of the core should provide a sequence of temperature changes of varying amplitudes (see Fig. 2.13 A, B).

The main problem involved in this technique is that of time-calibration (Mörner, 1972). The annual accumulation layers become progressively less visible at depth, in that they are squeezed out by compaction. Thus certain theoretical assumptions have to be made in the interpretation of the lower parts of these cores. In general, however, the results both from Byrd Station, and from Camp Century, have tied in well both with each other, and with other lines of evidence.

Geomorphic and pedologic evidence for environmental change

Although the data provided by ocean floor stratigraphy and by palaeo-ecological examination of terrestrial sequences have proved them to be two of the most profitable ways of reconstructing Pleistocene conditions, the importance of the evidence provided by fossil landforms and soils should not be forgotten. It is not possible here to go into detail about the relationships between landforms and climate, and fossil landforms and fossil climates, but there are certain landforms which can give relatively precise information about past environments (Table 1.3). Under cold conditions, with permafrost, for example, various types of patterned ground and ice-cored mounds, such as pingos will develop. As permafrost distribution is closely related to mean annual temperatures (see p. 42), former mean annual temperatures can be inferred from the distribution of fossil patterned ground and pingos. Likewise, the presence of glacial cirques can be used to infer the positions of former snowlines, which are themselves climatically controlled. The median level of a cirque floor tends to be at, or just above, the local snowline, so that the lowest cirque floor of a group of

Table 1.3
Some quantitative and semi-quantitative geomorphic indicators of environmental change

Landform	Indicative of	Example
Pingos, palsen, ice-wedge casts, giant polygons	Permafrost, and through that, of negative mean annual temperatures	Williams (1975), Great Britain
Cirques	Temperatures through their relationship to snowlines	Kaiser (1969), European mountains
Closed lake basins	Precipitation levels associated with ancient shoreline formation	Dury (1973), New South Wales
Fossil dunes of continental interiors	Former wind directions and precipitation levels	Saharan margins, Grove and Warren (1968)
Tufa mounds	Higher groundwater levels and wetter conditions	Butzer and Hansen (1968), Kurkur Oasis
Caves	Alternations of solution (humidity) and aeolian deposition etc. (aridity)	NW Botswana (Cooke, 1975)
Angular screes	Frost action with the presence of some moisture	McBurney and Hey (1955), Libya
Misfit valley meanders	Higher discharge levels which can be determined from meander geometry	Worldwide (Dury, 1965)
Aeolian-fluted bedrock	Aridity and wind direction	Massachusetts, Rhode Island, Wyoming (Flint, 1971)
Oriented and shaped deflation basins	Deflation under a limited vegetation cover	De Ploey (1965), Congo Basin

contemporaneous cirques will give a close approximation of the local snow-line. Thus the height of Pleistocene snowlines can be compared with present-day snowlines, and, by a knowledge of lapse rates, some estimate of temperature change can be obtained. In warmer areas landforms can also be used to reconstruct past climatic conditions. For example, as we shall see in more detail later, large continental sand dunes only develop over wide areas where precipitation levels are below about 100–200 mm. Above that figure sand movement is drastically reduced by the development of an extensive vegetation cover. Thus if fossil sand dunes are currently found in areas of high rainfall, it tends to suggest that rainfall levels have increased since the dunes were formed. Conversely, the presence of extensive fossil lake shorelines may be used to infer hydrological changes from wet to dry, and attempts have been made on the basis of old lake volumes to estimate former precipitation levels (see p. 46).

Soils too may be used profitably in Pleistocene studies. The development of a soil, whose duration depends on the nature and chemistry of the sediments, the climate, the character of the fauna and flora, and the balance between erosion and deposition, requires considerable time. Times of soil formation tend to be times of relative geomorphic stability. Thus thick palaeosols in a sequence of loess, dune sand, or alluvium may give important evidence for a halt to deposition, and a change to a period of stability. In the case of sand dunes, to take one example, the stability may result from a period of increased vegetation cover brought about by a phase of greater precipitation. Moreover, within a complex depositional sequence the character of the palaeosols themselves may change, and features in the soils such as gleying, carbonate accumulation, the snail fauna, degree of leaching, and frost structures may be used to assess environmental changes (see Chaline, 1972, pp. 44 *et seq.*, and Kukla, 1975). However, it needs to be appreciated that the formation of soils and most landforms results from a wide range of factors, of which climate forms but one group. Climate itself is also extremely complex. The problems of interpretation that these considerations present can be examined through a study of river terraces. They may sometimes form as a result of non-climatic causes, such as tectonic change, sea-level change, glacial invasion of catchment areas, and so forth. However, even if one can eliminate non-climatic causes, it is difficult to draw precise inferences as to climate from alluvial stratigraphy within terraces because of the variety of possible climatic influences: amount of precipitation, distribution of precipitation throughout the year, mean and seasonal temperature, and other climatic variables. Moreover, the response of a stream, in terms of load and discharge, to changes in such climatic variables, will be influenced by vegetation cover, slope angle, the range of the altitude of the basin, and other circumstances. Thus a change in any one climatic factor might, within one area, lead to different responses in different streams, and even in different segments of a single stream. Consequently, extreme care needs to be exercised in the utilization of landforms such as terraces for the reconstruction of past climates and environments.

The era before the Pleistocene ice
To appreciate the role which Pleistocene environmental changes, reconstructed on the basis of such techniques as those already described, have played in the history of the Earth and of man, and to grasp the magnitude of the changes involved, it is necessary to have some regard for the environmental conditions pertaining in the preceding period—the Tertiary (see Table 1.4).

It is, it must be pointed out, extremely difficult to make any division which is both logical and rigid between the Pleistocene and the last period of the preceding Tertiary period, the Pliocene. Indeed, the Villafranchian (the earliest unit of the European Pleistocene), together with its marine equivalent, the Calabrian, have only been assigned to the Pleistocene rather than to the Pliocene since 1948.

On faunal grounds the Pleistocene has come to be regarded as the time when many of the modern genera, including elephant, camel, horse, and wild cattle, first appeared. Attempts have also been made to place the boundary between the Upper Pliocene and the Villafranchian of the Lower

Table 1.4

Subdivisions of the Tertiary Era

	Date of beginning in millions of years
Pleistocene (or Quaternary)	1·8
Pliocene	5·5
Miocene	22·5
Oligocene	36
Eocene	53·5
Palaeocene	65

(After Berggren, *Nature*, 1969)

Pleistocene by using tectonic breaks in the stratigraphic succession, though this provides a generally inadequate and unusable correlation basis, except on a very local scale. In Britain the division between the Pliocene and the Pleistocene is placed at the boundary between the Coralline Crag and the Red Crag of East Anglia. At this point there is a relatively clear stratigraphic break, a marked increase in the proportion of modern forms of marine Mollusca, and of Mollusca of northern aspect, and the first arrival (in the Red Crag) of elephant and horse.

In Europe the base of the Pleistocene Series and of the Quaternary System have been set by the appearance in Late Cainozoic sediments occurring in various parts of Italy of a cold-water marine fauna (the Calabrian) differing from the underlying Pliocene fauna. The newer fauna is characterized by the appearance of a dozen species of North Atlantic molluscs, and by certain Foraminifera. In northern Italy the marine strata of the Calabrian grade into Villafranchian and Upper Villafranchian continental sediments containing a distinctive mammal fauna (Emiliani and Flint, 1963).

An alternative way of placing the boundary is to do so on climatic grounds. In essence some people would place the line according to the first point where there is evidence of glaciation, and of rapid and relatively sudden temperature depression. Recent work involving some of the new techniques outlined above, including potassium–argon dating, Foraminiferal studies, the examination of volcanic lava structures, and the interpretation of deep-sea cores, suggest, however, that the former belief that glaciation was confined to the Pleistocene, and was not characteristic of the Tertiary, must now be discarded. It seems clear that glaciation was initiated in some areas in the middle Tertiary, and this has led R. F. Flint (1972, p. 2) to remark that 'perhaps the most stirring impression produced by recent great advances in the study of the Quaternary period is that the Quaternary itself is losing its classical identity'. Previously it had been widely believed that the period from the Triassic to the Tertiary was a lengthy span when ice sheets and glaciers were absent, and when climatic fluctuations were less frequent and less severe than they were to be in the Pleistocene. Bandy (1968), however, has written that 'the magnitude of the planktonic faunal changes indicates that the paleo-oceanographic changes of the later Miocene and the middle Pliocene are almost as great as those of the classic Quaternary.' By a study of *Globigerina* types in ocean cores he found evidence for a major expansion of polar faunas no less than between 10 and 11 million years ago. This expansion was followed, he suggested, by another mid-Pliocene expansion between 5 to 7 million years ago, and the classic Pleistocene expansion 3 million years ago.

Equally, some of the lithified glacial *tills* (non-sorted and non-stratified sediments carried or deposited by a glacier) interbedded with volcanic lavas in the White River Valley area of Alaska, have been dated at nine to ten million years old, and similar methods, combined with a study of the ice-berg-rafted debris in Southern Ocean deep-sea cores, suggest that the east Antarctic glaciers reached a full-bodied stage somewhat before five million years ago. Another core from the area has led to the even more striking conclusion that glacial conditions may have been present in the Eocene (Geitzenauer, 1968), for a coexistence of micro-fossils of Eocene age, and of sediments of glacial type, has been proved. In eastern Antarctica ice-field initiation dates back to the Lower Miocene (Drewry, 1975).

Confirmatory evidence of Eocene glaciation in Antarctica, with all that this implies for world sea-levels and world climate, is provided by the nature of dated volcanic materials on that continent. Volcanoes that have erupted beneath an ice sheet display a suite of textural and structural characteristics that are especially distinctive in volcanoes composed of basaltic lava, and such materials have been found in Antarctica back to the Eocene. Similar evidence implies also that glaciation may well have been uninterrupted until the present in this South Polar region (Le Masurier, 1972).

All these dates for the onset of ice caps and glaciation are very considerably earlier than the classic dates of around one million,[1] to one and a half million years formerly given for the start of a 'climatic' Pleistocene. Now the Pleistocene seems to be accorded a length of the order of two and a half to three million years, but a further discussion of this issue must wait until the next chapter.

Even though this new evidence has greatly altered our view of the Tertiary era there is probably a considerable amount of truth in the oft stated idea that in general temperatures did show a tendency to decline in many parts of the world during the course of the Tertiary. On the basis of deep-sea core studies, Emiliani (1961), a pioneer of the technique, has suggested that there was a broad temperature decrease amounting to about 8 °C for the middle latitudes during the Tertiary. A similar temperature depression is indicated in the Equatorial Pacific, but on the basis of land

[1] Some authorities gave even shorter duration to the Pleistocene. Zeuner, for example, in 1959 gave an estimate of only 600 000 years.

Table 1.5

Tertiary mean annual temperatures (°C)

	NW. Europe	W.U.S.A.	Pacific Coast of North America
Recent	—	—	10
Pliocene	14–10	8–5	12
Miocene	19–16	14–9	18–11
Oligocene	20–18	18–14	20–18·5
Eocene	22–20	25–18	25–18·5

(From data in Butzer, 1972)

flora studies somewhat greater changes than these have been proposed for the western United States between 40 and 50 °N and the picture for other areas is the same (Table 1.5). The pattern of temperature decline for Pacific Ocean water, with a progressive southward shift in isotherms as the climate became cooler, is illustrated in Fig. 1.5.

The North Atlantic region in the Early Tertiary was characterized by a widespread, tropical rainforest vegetation (Pennington, 1969, Chapter 1). In the London–Hampshire Basin of southern England, for example, the Malaysian *Nypa* palm and mangrove (*Avicennia*) swamp was present, while in the north German plain, in brown coal deposits, leaves of *Pandanus*

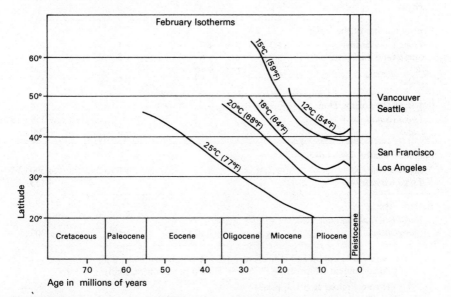

FIG. 1.5. Tertiary isotherms for Pacific Ocean water. During Tertiary times the isotherms for the water of the Pacific Ocean gradually shifted to the south as the climate became colder. About 35 000 000 years ago, for instance, the 20 °C isotherm was at the latitude of Seattle, but it moved to Baja California as the cold advanced from the north (after Kurtén, 1972, p. 28).

Table 1.6
Differences in temperature and annual precipitation between Reuverian B (the end of the Pliocene) and the present estimated from climatic values based upon the current distribution of certain plants

Region and genus	1	2	3	4
NW. Germany and Netherlands				
Liquidambar	+3 to 4		= or +1 to 2	
Nyssa near coastline				
edge of Mittlegebirge	+3			
Tsuga (*canadensis*)	+2			+300
Carya and *Liriodendron*				
near coastline edge of				
Mittlegebirge	+3			
Ilex				
Fagus				
Central and east Poland, Lithuania and White Russia				
Liquidambar	+5	+5	+3 to 4	+350
Nyssa	+4 to 5	+1	+3 to 4	+250
Tsuga (*canadensis*)	+2 to 3	+3	+1	+400
Carya	+3			
Ilex		+5 to 6		+150
Fagus		+4	+3	
Central reaches of the Volga				
Tsuga (*canadensis*)	+2 to 3	+3	+1	400
Carya (and *Liriodendron*?)	+3	+6	+5	350
Ilex		+15		250
Fagus		+12	+5 to 6	200
Lower reaches of the Don				
Tsuga (*canadensis*)				+350
Fagus		+7	+3	+200
Bulgaria—central lowlands				
Liquidambar	+3	+1	+1 to 2	+300
Tsuga (*canadensis*)				+150
Eastern Siberia				
Tsuga heterophylla and				
mertensiana		+10 to 15	+10	+200

Notes: 1 = July mean temperature (°C), 2 = January mean temperature (°C), 3 = Mean annual temperature (°C), 4 = annual precipitation (mm).

(From Frenzel B. 1973, p. 89)

(screw pine), a tropical plant with stilt roots have been found. Deposits from the Bovey Tracey fault basin in western England, which are probably of pre-Miocene age, contain many tropical forest plants, including *Osmunda*, *Calamus*, *Ficus*, *Symplocos* and *Laurus*. Likewise, in the early Tertiary, Arctic islands such as Greenland, Spitzbergen, and even Grinnell Land (81° 45′N.) were forest clad. However, by Pliocene times the tropical vegetation of the North Atlantic region was being replaced by a largely deciduous, warm temperate flora including *Sciadopitys*, *Tsuga*, *Sequoia*, *Taxodium*, and *Carya*, some of which were to be eradicated by the intense cold of the Pleistocene itself (Montford, 1970). The last appearance of palms north of the Alps in Europe is in the late Miocene flora at Lake Constance.

Immediately prior to the Ice Age, however, in a period which the Dutch call Reuverian B, the climate of the present temperate latitudes of the northern hemisphere seems to have favoured the development of woodland. At that time varied woodlands extended from the Atlantic coasts to the Sea of Japan, a picture that has not been repeated since (Frenzel, 1973). Over wide reaches of the present warm and cool temperate latitudes, temperature and rainfall conditions were particularly favourable, and resembled those of a subtropical climate. Some tentative quantitative estimates based on the analysis of present and past distributions of certain species or genera are shown in Table 1.6. On average, in central and eastern Europe, temperatures were probably 3 to 5 °C higher than they are now, and precipitations levels were several hundred millimetres higher.

In Australia a comparable sequence of decline of Tertiary temperatures has been proposed, based on the former, great extent of *Araucaria* and *Agathis* into Tasmania. At the present time they are limited to Queensland and the warmer parts of Australia. However, rainfall decline may have been equally important in such changes of floral distributions, and Gentilli (1961) has said that 'areas that now receive 12·5 cm of rain a year must then have received at least 125 cm with no rainy season. If there was even a short dry season the annual rainfall must have reached some 200 or 250 cm for these laurisilvae to grow as they did.' Large trees then existed in the Lake Eyre Basin and other stretches of the 'dead heart' of Australia (Gentilli, 1961; Gill, 1961).

The widespread nature of warm, wet conditions in certain parts of the world in Tertiary times had diverse environmental effects. Deep-weathering profiles were produced over wide areas in mid-latitudes, in the form of laterites, and silicified layers called silcretes. Limestone areas were subjected to intense solutional processes, and rocks were rotted so that they were to be particularly susceptible to glacial erosion in the Pleistocene.

Selected Reading
The literature on environmental change is massive. This list of selected reading refers to those papers and books which are either particularly accessible, or particularly relevant to certain themes.

In recent years some excellent general reviews on environmental change have been published. Particularly valuable is R. F. Flint's (1971), *Glacial and Quaternary Geology*. It contains a lengthy multi-lingual bibliography. Also useful, not least for its information on low-latitudes, is K. W. Butzer's (1972), *Environment and Archaeology: an ecological approach to prehistory*. K. K. Turekian (1971) has edited *The Late Cenozoic glacial ages*, and this contains a series of essays concerned primarily with recent developments. J. Chaline's (1972), *Le Quaternaire* is a valuable general French survey.

Flint has a lengthy section on the development of ideas on environmental change. A scholarly work on British ideas is G. L. Davies's (1967), *The Earth in decay* (Macdonald), whilst Toulmin and Goodfield's (1965), *The discovery of time* (Pelican) provides a stimulating discussion of how man's concept of time has developed.

The new chronological techniques, primarily radiometric, are discussed in W. W. Bishop and J. A. Miller (eds.) (1972), *The calibration of hominoid evolution* (Chatto and Windus), and in Olsson, I. U. (ed.) (1970), *Radiocarbon variations and absolute chronology* (Wiley). Some other techniques including pollen-analysis (palynology) are discussed in R. G. West (1972), *Pleistocene geology and biology*. A stimulating analysis of dating as the basis of chronology is in C. Vita-Finzi (1973), *Recent earth history*.

Discussion of the climatic conditions of the Tertiary, and their stratigraphic relations, appear in a number of works. Among the more valuable general works are: Berggren, W. A. (1969), 'Cainozoic Stratigraphic, planktonic foraminiferal zonation and the radiometric time-scale', *Nature*, 224, 1072–5; H. M. Montford (1970), 'The terrestrial environment during Upper Cretaceous and Tertiary times', *Proceedings Geologists' Association of London*, 81, 181–204.

Two recent publications which give a good impression of the way that techniques are evolving and thoughts changing are the World Meteorological Organization's (1975) *Proceedings of the WMO/IAMAP symposium on long-term climatic fluctuations, Norwich* (WMO, Geneva) and the United States Committee for the Global Atmospheric Research Program's (1975) booklet *Understanding climatic change, a program for action*. Likewise, the immensely important results that have come from the use of ocean cores are illustrated in R. M. Cline and J. D. Hays (1976) (eds) 'Investigation of Late Quaternary Paleoceanography and paleochimatology' *Geological Society of America, Memoir*, 145.

2 The Chronology and Nature of the Pleistocene

> The large number of glacial stages indicated may seem
> surprising, but may perhaps be the less so if the problem
> of estimating the number of glaciations from evidence
> on continents is contrasted with the problem of estimat-
> ing the number of glaciations from evidence in deep-sea
> sediments. The first may be compared in complexity to
> estimating the number of times the blackboard has been
> erased; while the second may be compared to finding the
> number of times the wall has been painted.
>
> M. Ewing (1971, p. 572)

Introduction

The Pleistocene did not consist of just one great ice age, but was composed
of alternations of great cold (glacials, stadials), with stages of relatively
greater warmth (interglacials, interstadials).

The expansion and thickening of the ice sheets and glaciers in the glacial
stages led to the erosion of underlying rock and to the transport of large
quantities of debris over long distances. This debris, which is given a large
variety of names, including till and boulder clay, is a highly characteristic,
normally ill-sorted combination of boulders, sands, and clays. The debris
frequently contains rocks derived from some hundreds of kilometres away.
These are called erratics and may attain a considerable size. In eastern
England, near Ely and also on the Norfolk coast, some of the erratics
produced by glacial erosion of the chalk attain colossal dimensions, some
being 400–600 m long and up to 50 m thick. In interglacial periods, when
conditions became warmer, like those of the present time, the ice retreated
and left moraines and other related glacial and fluvio-glacial landforms
and deposits. These were then weathered. Other sediments might also
accumulate on top of the glacial deposits, and these contain characteristic
faunal and floral remains (see p. 56). In another glacial period such
sediments might themselves become covered by boulder clay. Classic interpretations of the history of the Ice Ages or the Pleistocene have been based
on a study of the extent and character of these alternations of glacial and
interglacial deposits on land.

Although Geikie and Ramsay established over a century ago that the
Pleistocene Ice Age was composed of multiple glaciations, and in spite of

the great deal of work that is being devoted to establishing the duration of the Pleistocene epoch at the present time, there is still a marked degree of controversy over the number of glaciations, stadials, interglacials, and inter-stadials. This exists partly because of the problem of definition of these events, a matter discussed further on page 53. There is also a lack of agreement with regard to correlations of events between different areas, and there is still no universally agreed idea, as mentioned in the previous chapter, on the date of the Pliocene-Pleistocene boundary. Nevertheless, the great increase in the use of new dating techniques, and of deep-sea core evidence, have enabled some statements to be made with a greater degree of confidence than hitherto.

The length of the Pleistocene

There exists a considerable range of views as to the duration of the Pleistocene epoch, and also as to its definition. In recent years there has been a tendency to place the boundary between the Pliocene and the Pleistocene on faunal grounds (such as the appearance of certain Foraminifera and the extinction of Discoasteridae) at about 2 million years B.P., and on climatic grounds (the marked appearance of mid-latitude as opposed to polar glaciers) at 2·5 to 3·0 million years B.P. (i.e. at the end of the ·Mammoth polarity event). This is a very considerable increase on former estimates, but there appears to be a discrepancy of around half a million years between the climatic and faunal boundaries. Table 2.1 gives a list of both faunal and climatic dates from a very wide range of world environments. Given the present state of knowledge, and the discrepancies that exist, it is probably reasonable to state that the duration of the Pleistocene has been around 2–3 million years.

The divisions of the Pleistocene

The use of modern techniques of investigation of the lengthy stratigraphic record provided by the sea-floors has led to new views being expressed on the length and frequency of glacial and interglacial episodes. In the last decade Emiliani (1968), Kennett (1970), Kent, and other workers (1971) have established that there have been rather more cycles than was previously thought possible. Although they do not agree on the exact number of cycles, Emiliani, for example, favouring 20 glacial cycles in the last million years, and Kent and his co-workers 16 glacial phases in the last 2·5 million years, these new data, with their suggestions of a large number of glacial-interglacial cycles give a very different picture from the classic four-fold glacial sequence proposed between 1901 and 1909 by Albrecht Penck and E. Brückner in their three volume work, *Die Alpen im Eiszeitalter*. By studying what was left of old moraines, especially terminal moraines, they were convinced that there had been four great glaciations of different intensities in the Alpine region. They also proved, from plant remains at Hötting

Table 2.1

Faunal and climatic dates for the start of the Pleistocene

Source	Location	Evidence	Date (m. yrs. B.P.)
Faunal dates			
Leakey (1965)	Olduvai (Tanzania)	K/AR dating of upper Villafranchian fauna	more than 1·75
Glass *et al.*, (1967)	Deep-sea	Coiling reversal in Foraminifera	2·1
Glass *et al.*, (1967)	Deep-sea	Discoasteridae [1] die out	2·0–1·8
Zagwijn (1974)	North Sea Basin	Faunal extinctions	2·5
Climatic dates			
McDougall & Wensink (1966)	Iceland	Tillite/basalt	3·0
Mathews and Curtis (1966)	New Zealand	Tillite/basalt	2·47
McDougall & Stipp (1968)	Sierra Nevada	Tillite/basalt	2·7–3·1
McDougall & Stipp (1968)	Taylor Valley	Tillite/basalt	2·7
Opdyke *et al.*, (1966)	Southern Ocean cores	Ice-rafted debris	2·5
Mercer (1969)	Argentina	Tillite/basalt	2

[1] *Discoasteridae* were a distinctive group of planktonic organisms, which secreted six-rayed, star-shaped skeletal elements known as *Discoasters*.

and other sites, that some of the intervening periods were fairly mild. They further demonstrated the correspondence between these four glaciations and the successive gravel terraces of the Rhine and other rivers. They named the four glacial periods after valleys in which evidence of their existence occurred, the Günz, Mindel, Riss, and Würm.

The Penck and Brückner model was widely accepted, and almost attained the status of a law. Evidence for numerous additional glaciations obtained by other workers was generally forced into their scheme, and they were assigned the secondary role of substages or stadia within the four major glaciations. In many respects, however, the Alpine sequence was an unfortunate base for long distance correlation, in that the Alpine area is not ideal for Quaternary stratigraphic studies. There is a relative lack of organic deposits, there is some difficulty in relating such deposits to the glacial deposits, there are considerable possibilities for confusion resulting from earth movements, and complications are introduced by the separation of key areas by mountain chains. As Sparks and West (1972) proclaim, 'The eradication of the Alpine nomenclature, which should never have been applied widely in the first place, has proved and indeed is still proving a Herculean task.' Also proving to be a Herculean task is the establishment of an acceptable stratigraphic scheme to replace the Penck and Brückner model. Pilbeam (1975, p. 819) has remarked that

A wide variety of stratigraphic schemes have been proposed for the last 2 to 3 million years . . . None of these . . . agrees precisely with any of the others; most are probably about equally acceptable; and, no doubt, none is absolutely correct. However, all are useful advances beyond the four glacial-pluvial schemes still utilized so widely in many anthropological textbooks.

Thus there is still a considerable degree of latitude in current ideas on the dates and lengths of the major events of the Pleistocene. The antipenultimate major glaciation in the Alps, for instance, the Mindel of Penck and Brückner, was even in the 1960s assigned by different authorities a duration that varied by a factor of twenty, and a beginning date that differed by a factor of seven! Ericson and Wollin (1968) for example, giving a date of 1 400 000–915 000 years B.P., and Emiliani (1961) of 200 000–175 000 B.P. Figure 2.1 brings this out further. The 'long' temperature curve of Ericson and Wollin (Fig. 2.1A), derived from a study of Foraminifera distributions in ocean cores, is very different from the 'short' one obtained by Emiliani (Fig. 2.1B) on the basis of the oxygen-isotope ratios of Foraminiferal tests from cores. The Emiliani curve includes nine warm periods in the past 450 000 years, and displays a large number of short-term fluctuations that are not shown in the Ericson and Wollin curve.

This difference in interpretation of the deep-sea core record, and the

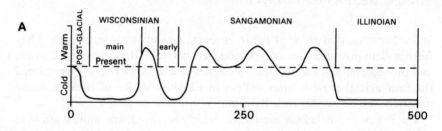

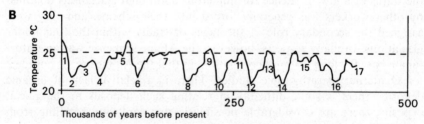

FIG. 2.1. Two different climatic curves for the last 500 000 years.
    A. The generalized climate curve based on data of Foraminifera in deep-sea cores as determined by Ericson and Wollin (1968, Fig. 7).
    B. The palaeotemperature curve based on oxygen-isotope ratios as determined by Emiliani (1966, Fig. 5). The even numbers represent cold phases, the odd represent warmer.

problems that arise when attempting correlations with imperfectly dated terrestrial sequences is further exemplified in Fig. 2.2 which shows three long deep-sea cores. One interpretation (C) compresses the 'classic' American glacial and interglacial episodes into about 800 000 years whereas the other two (A, B) suggest a much longer duration. Discrepancies

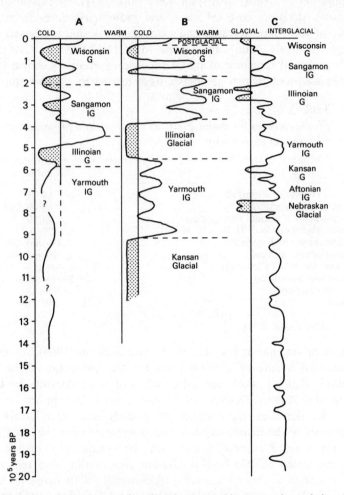

Fig. 2.2. Three long deep-sea core palaeoclimatic curves with attempted correlations with the American terrestrial sequence of glacials and interglacials.

    A. Generalized curve from 12 south Pacific cores based on variations in Foraminiferal faunas (after Kennett, 1970, Fig. 8).

    B. Generalized climate curve based on variations in the frequency of *Globorotalia menardii* Foraminifera in cores from the Atlantic Ocean (after Wollin *et al.*, 1971, Fig. 7).

    C. Generalized climate curve based on frequency of ice-rafted debris in cores from the North Pacific (after Kent *et al.*, 1971, Fig. 8).

The glacial peaks have been emphasized with shading by the author.

arise because of the non-parallel nature of the curves, and from the different interpretations that can be placed on different sizes and lengths of peaks.

There is thus considerable debate as to whether the classic glaciations occurred in a relatively short time-span or whether they were spread out. One supporter of a short time-span was Evans (1971). He examined insolation curves, deep-sea core evidence, and radiometric dates from Europe (notably of the Rhine terraces and associated volcanic materials) and assigned, a relatively young age to the Mindel (320 000 to 220 000 B.P.). Evans's proposed sequence of major Pleistocene events is shown in Table 2.2. European and British geologists have tended to support this shorter

Table 2.2

Proposed dates for major Pleistocene events by a proponent of a 'short' Pleistocene chronology

Event	Age in t.y. B.P.
End of Last (Weichsel, Würm) glaciation	10–11
Beginning of Last Glaciation	60–80
Eemian maximum of warmth (interglacial)	90–80
Penultimate glaciation (Saale II, Warthe, Riss II)	120–100
Fluctuating climate with warmth	c. 170 and 130
Fluctuating climate with cold	c. 160 and 140
Saale I, Drenthe, Riss I Glaciation	200 (or slightly less) to 180
Holstein/Hoxne Interglacial	220–200 (or slightly less)
Mindel/Elster glaciation	320–220
Interglacial	380–320

(After Evans, 1971)

time-scale or one like it (Cooke, 1973), and Shotton (1966), for example, has suggested a date of 275 000 years for the mid-point of the Mindel Glaciation. Kukla's work on palaeosols and loess deposits in Czechoslovakia and Austria (Kukla, 1975) also tended to support the 'short' model. He identified eight cycles of glacials and interglacials in the 700 000 years of the Brunhes epoch, and seventeen within the last 1 600 000 years. However, Richmond's work on the stratigraphy of the Rocky Mountains which utilized both potassium argon dating and tephrochronology to obtain a time framework (Richmond, 1970) came to different conclusions. He suggested that the Nebraskan Glaciation ended about 1·2 million years ago, and therefore endorsed the concept of a long chronology.

Given the present general state of knowledge it is impossible to come to any widely acceptable conclusion about the validity of these two different interpretations of the sequence of Pleistocene glacials and interglacials. The dating procedures for the terrestrial sequences are still inadequate, the curves obtained from deep-sea cores are susceptible to different interpretations, the terrestrial record is often incomplete, the record of the last glacia-

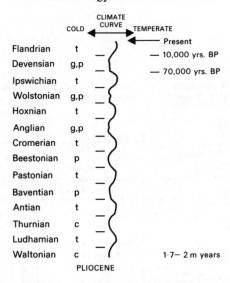

FIG. 2.3. The sequence of stages in the British Pleistocene in relation to climatic change. t = temperate, g = glacial, p = permafrost, and c = other evidence for cold conditions (after Sparks and West, 1972, Fig. 6.6).

tion suggests that the glacials themselves were complex events, and the terrestrial record left by smaller glacial phases could easily have been eradicated by succeeding larger glaciations. This dilemma is succinctly stated in the epigraph at the start of this chapter.

Moreover, in some areas cold phases may not have led to actual glaciation. This is, for example, the case in the Early Pleistocene of eastern England, where sediments in a borehole at Ludham indicate fluctuations from forest to a type of oceanic heath, and back to forest. This oceanic heath is interpreted as a result of climatic change for the worse, which probably produced glacial conditions in higher latitudes. Evidence for similar Lower Pleistocene climatic oscillations has also been found from boreholes in the Netherlands. The British sequence, obtained from work in East Anglia, and illustrating the 'pre-glacial' cold phases of the Early Pleistocene, is shown diagramatically in Fig. 2.3. At least seven cold phases have now been identified since the Pliocene, but it would appear that these cold phases were not severe enough for the production of glaciers until the beginning of the Anglian stage, when North Sea ice entered Norfolk to deposit the Cromer Till.

The Netherlands continental sequence (after Zagwijn) is illustrated in Fig. 2.4. While it cannot yet be directly correlated with the East Anglian sequence it shows a similar pattern of fluctuations with miscellaneous Early Pleistocene oscillations following on from the rather warm conditions of the Pliocene Reuverian (see p. 23). The palaeobotanical evidence for the early

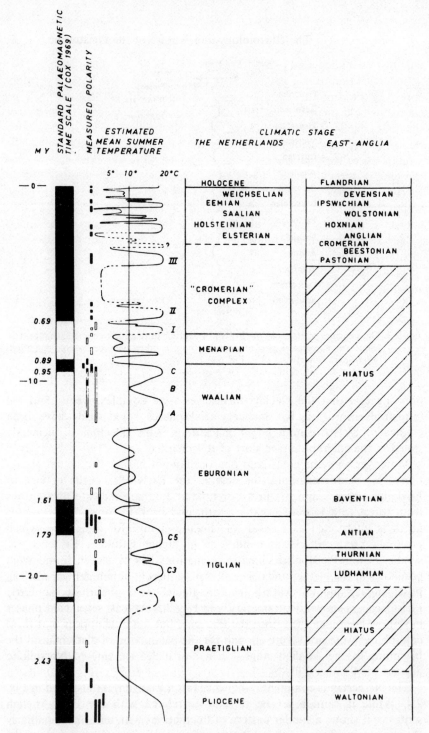

Fig. 2.4. Palaeoclimatic measurements, climatic curve, and stratigraphic climatic division of the Quarternary in the Netherlands (after Zagwijn, 1975, Fig. 8).

cold phases is good, and permafrost structures have also been identified, but actual glaciation in the Netherlands would only appear to have been widespread in the penultimate cold phase, the Saalian. In all there were at least six major cold phases revealed in the non-marine Pleistocene sediments of the Netherlands. Zagwijn (1975) believes that the climatic curve shows two important trends during the course of the Pleistocene. The first is that the amplitudes of the oscillations seem to increase. This was caused mainly by the temperatures of the cold phases becoming cooler, for temperatures in the warm phases remained fairly similar throughout the Pleistocene. The second trend that can be identified is that the frequency of the oscillations shows a distinct increase upwards, especially so after the Jaramillo palaeomagnetic event about 900 000 years ago. Such a trend has also been identified in some of the deep-sea cores (see, for example, the concentration of peaks during the last 600 000 to 900 000 years in the three cores illusrated in Fig. 2.2).

A somewhat similar picture has been put together for Central Europe by Segota (1966). By examining the miscellaneous evidence provided by floral studies he suggested the sequence of events illustrated in Fig. 2.5. Once

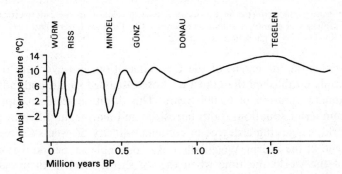

FIG. 2.5. Curve of Quaternary temperature changes in Central Europe (after Segota, 1966, Fig. 1).

again the concentration of the classic glaciations in the later part of the Pleistocene is apparent, the frequency of the fluctuations in the same period becomes evident, and the intensity of the cold in the classic glacial phases appears to be greater than in the 'pre-glacial' Pleistocene.

Another attempt at relating the classic cold phases or glacials of the Pleistocene to the whole, extended spread of the Pleistocene, is provided by an examination of the sequence from European Russia (Fig. 2.6). As in eastern England and the Netherlands it is apparent that the four major glaciations that have been identified cover only a portion of the total length of the Pleistocene, with the start of the Odessa Glacial, which is possibly to be correlated with the Günz, being dated at about 1·4 million years ago. Before that time there were miscellaneous fluctuations of climate.

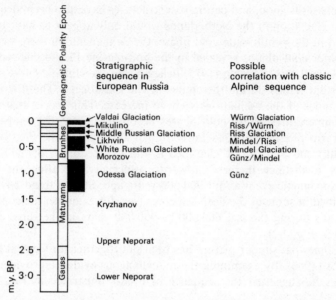

FIG. 2.6. Stratigraphic sequence in European Russia in relation to the radiometrically-dated sequence of geopolarity units as determined by Gromov (after Flint, 1971, Table 24-K). The classic glaciations are shaded.

The dates for the extent of the Last Glaciation, however, are relatively more firmly established than most of those discussed so far, and indicate an approximate duration of 60 000 years. This duration seems to apply to a large number of situations, both terrestrial and marine, in different parts of the World, suggesting a degree of contemporaneity of events (Table 2.3).

The end of the Würm–Weichselian–Wisconsin glacial period was in many parts of the world the time when the ice sheets reached their maximum extents, and conditions were at their most frigid. The last great interstadial ended around 23 000 years B.P. and after this time the ice sheets seem to have expanded markedly, and within about four or five thousand years they reached their maxima. Table 2.3 shows the various dates clustering around 18 000 B.P. that have been suggested for the maxima of the Last Glaciation.

Thereafter, deglaciation set in, though the process was not continuous, and the years up until the start of the Holocene were marked by small interstades and re-advances (stadia).

One can say that the Holocene or Recent[1] started about 10 000 years ago in that warming progressed very rapidly after that date. At around 9500 B.P., for instance, the eastern Greenland Ice Cap was retreating very fast—probably at a rate of 3 km/100 years. On the other hand, some warming

---

[1] An alternative name for the Holocene or Recent is 'Flandrian'. This has generally been a term utilized with respect to the transgression caused by the worldwide rise in post-glacial sea-level. It is now sometimes utilized with a broader connotation.

took place after a worldwide cold phase which was peaked at around 14 500–14 000 years B.P. and some authorities (see, for example, Mercer, 1972) believe that it might be desirable to place the Pleistocene–Holocene boundary at this point. This period of rapid warming and deglaciation can be identified in the oceanographic record (Kennett and Shackleton, 1975). Freshwater derived from melting ice is relatively impoverished in O^{18}. Thus marine waters diluted by glacial meltwater are isotopically more negative than undiluted oceanic waters. There is a massive anomaly in the isotopic composition of Gulf of Mexico cores between 15 000 and 12 000 B.P. This is related to massive inpourings of glacial meltwater into the Gulf of Mexico via the Mississippi system which then drained the melting Laurentide ice sheet. Surface water salinities were evidently reduced by around 10 per cent throughout the western Gulf.

Table 2.3
Dates for the start, maximum, and end of the Last Glaciation

Source	Location	Start	(yrs. B.P.) Maximum	End
Flint (1971), and Segota (1966)	Central Europe and Germany	70 000	18 000–20 000	10 000
Heusser (1961), and Mercer (1972)	North America	—	18 000–21 500	9 000–10 000
Epstein *et al.* (1970)	Antarctic	75 000	17 000	11 000
Dansgaard *et al.* (1969)	Greenland ice core	73 000	—	10 000
Ericson *et al.* (1961)	Ocean cores	60 000	—	11 000
Mercer (1972)	New Zealand	—	18 000	—
„	Chile	—	19 400	—
Kind (1972)	U.S.S.R.	—	—	10 400
Rona and Emiliani (1969)	Caribbean core	—	17 000	—

Terminology in different regions

In view of the already stated uncertainties of major events like the Mindel Glaciation of the Alps (see p. 28), correlation of glacial and interglacial phases in different regions is a hazardous procedure given the inadequacies of modern dating methods, and the fragmentary nature of the stratigraphic evidence. However, in Table 2.4 a list of major events is given for different regions from the northern hemisphere, but deliberately no direct correlations are attempted. Its purpose is to enable the reader to obtain some meaning from the local terminology which is utilized in various parts of this book. Nevertheless, radiocarbon dates are now available which allow the last glaciations (Weichselian, Devensian, Würm, Valdai, Wisconsin) to be correlated, and many workers have suggested that the last four cold events (equivalent to the four classic glacials of the Alpine sequence) may also be

Table 2.4

Sequences of Pleistocene phases in the northern hemisphere

	Rhine Estuary[1]	Britain[2]	Alpine foreland[3]	European Russia[4]	North America[5]
	Holocene[6] ———→				
Glacial Pleistocene	WEICHSELIAN	DEVENSIAN	WÜRM	VALDAI	WISCONSIN
	Eemian	Ipswichian	Riss-Würm	Mikulino	Sangamon
	SAALIAN	WOLSTONIAN	RISS	MIDDLE RUSSIAN	ILLINOIAN
	Holsteinian	Hoxnian	Great Interglacial	Likhvin	Yarmouth
	ELSTERIAN	ANGLIAN	MINDEL	WHITE RUSSIAN	KANSAN
	Cromerian	Cromerian	Günz-Mindel	Morozov	Aftonian
	MENAPIAN	BEESTONIAN	GÜNZ	ODESSA	NEBRASKAN
'Pre-glacial' Pleistocene	Waalian	Pastonian	Donau-Günz	Kryzhanov	
	EBURONIAN	BAVENTIAN	DONAU		
	Tiglian	Antian			
	PRETIGLIAN	THURNIAN			
		Ludhamian			
		WALTONIAN			
Pliocene					

Notes
[1] From Zagwijn (1975).
[2] From Sparks and West (1972).
[3] From Penck and Brückner (1909) and others.
[4] From Flint (1971).
[5] From Flint (1971).
[6] Interglacials are in lower case, Glacials in capitals.

correlated, though the means whereby this can be done with any degree of certainty are sparse. It is certain that correlations become increasingly hazardous as one moves back through the classic 'glacial' Pleistocene into the 'pre-glacial' Pleistocene. The main problem is one of dating, and as Vita-Finzi (1973) has remarked, 'When correlation precedes dating it is difficult to argue about contemporaneity, let alone about age differences.'

The changing extent of glaciers and ice caps

The Pleistocene glaciers did not show precisely the same spread in each of the glacial periods. In Europe the Riss/Saale glaciation is regarded as representing the maximum spread of ice, and although the Illinoian (Table 2.4) probably represents the maximum spread of ice in North America, the preceding Kansan glaciation extended further in the west-central part of America. At their maximum extent the glaciers probably covered 47·14 mkm², which is rather more than their extent during the Last Glaciation (40·30 mkm²) but very much more than their present extent, which is about 15 mkm². In other words during the height of the Pleistocene glaciations, ice covered around three times as great an area as it does today (Embleton and King, 1967). However, both the Antarctic and Greenland ice sheets, because of the restrictive effects of iceberg-calving into deep ocean water, differed little in area from their present sizes, though they were much thicker. The greatest reductions in the glaciated area have taken place through the melting of the North American Laurentide and the Scandinavian ice sheets, which have now lost 99 per cent of their former maximum bulk. In terms of volume, although precise assessment is difficult, the Riss/Saale/Illinoian glaciation had a volume of 84–99 mkm³ compared to 28–35 mkm³ at the present time (Table 2.5).

Table 2.5

The former and present extent of glaciated areas

	Maximum Extent (mkm²)	Last Glaciation	Present day (km²)
Antarctica	13·20	13·20	12 650 000
Laurentide	13·79	12·74 ⎞	230 250
N. American Cordillera	2·50	2·20 ⎠	
Siberian	3·73	1·56	—
Scandinavian	6·67	4·09	5 000
Greenland	2·16	2·16	1 800 000
N. Hemisphere besides above	4·07	3·45	—
S. Hemisphere excluding Antarctica	1·02	0·90	26 000
Total (mkm²)	47·14	40·30	14·97

(From Embleton & King, 1967)

The extension of the glaciers and ice sheets in key areas

America‹

During the maximum glacial phases of the Pleistocene, including the late Wisconsin maximum around 18 000 years ago, ice was continuous or nearly so across North America from Atlantic to Pacific, and was composed of two main bodies, the Cordilleran glaciers associated with the Coastal Ranges and the Rockies, and the great Laurentide sheet (see Wright and Frey, 1965). The former were most extensive in the mountains of British Columbia, and diminished both northward into Alaska and Yukon, and southward through the western United States. The southern limit of continuous ice was south of the Canada-U.S.A. border, down to the Columbia River and the Columbia Plateau. South of this there were numerous localized ice caps and glaciers, and notably in the Sierra Nevadas the thickest ice, in the lee of the Coastal Ranges, was as much as 2300 m deep. The Laurentide ice sheet, which, as already mentioned was with the Scandinavian ice sheet largely responsible for the great difference in world glacial areas between glacial phases and interglacials, reached its extreme extent in the Ohio-Mississippi basin at latitude 39° in Wisconsin times, and 36° 40′ in Illinoian times. It extended to approximately the present positions of St. Louis (Missouri) and Kansas City (Kansas), but west of this area the southern ice margin trended north-westwards, leaving western Nebraska and western South Dakota largely ice-free. On the basis of post-glacial isostatic readjustment it seems likely that the thickest ice occurred over Hudson Bay, and attained a thickness of around 3300 m.

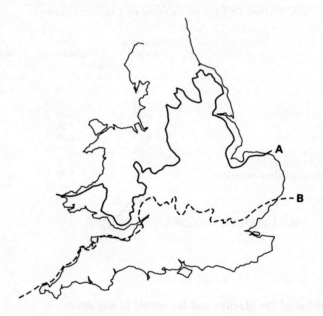

Fig. 2.7. Glacial limits in England and Wales.
    A. The limit of glaciation during the Devensian. The maximum in Wales is still the subject of some discussion.
    B. The maximum extent of glaciation in southern England as generally accepted.

The British Isles

The British Isles ice sheets (with an area of about 370 000 km² during the Last Glaciation) merged with those of Scandinavia, but also possessed large local centres of ice dispersal, including the Scottish Highlands, the Lake District, the Southern Uplands, the Pennines, the Welsh mountains, and various mountains in Ireland including those of Connemara and Donegal in the west, and those of Kerry and Wicklow in the south and east respectively. The extent of the various glaciations (Fig. 2.7), notably in South Wales

(Bowen, 1973) and in Wessex, is the subject of some dispute, and the debate over the southern limits of glaciation has been renewed with some vigour during the early 1970s.

From work in North Devon and the Isles of Scilly it became clear that an ice sheet at some time reached the north coast of the south-western peninsula. Till has been identified on the Isle of Lundy, at Fremington near Barnstaple in Devon, and in the Scilly Isles, and possible glacial overflow channels have been described from near Lynmouth and Hartland Quay. New motorway sections revealed probable boulder clay in the area south of Bristol, indicating the possibility that westward moving ice crossed the Severn. Information such as this has been used to prepare the line of the southern limit of glaciation used in Fig. 2.7.

Occasionally it has been postulated that glaciation was even more extensive than that map would suggest (see, for example, Kellaway, 1971), but the bulk of the evidence is sparse, and usually open to an alternative explanation, and studies of river gravels in the basins of the Test, Avon, and Axe (Green, 1973 and 1974) fail to show that glaciers ever introduced erratic materials into them.

The limits of glaciation in the Devensian (Last Glacial) are better known, and they appear to have been further north than during the earlier glaciations, so that in East Anglia, for example, the ice only just touched the Norfolk coast at Hunstanton. In previous glaciations it reached down to just to the north of London.

Europe and Asia

In mainland Europe and Asia there were three main centres of ice, the Alpine, the Siberian, and the Scandinavian.

The Alpine glaciers covered an area that has been estimated at 150 000 km² and reached down to altitudes of 500 m on the north, and 100 m on the south side. The ice may have been over 1500 m thick in places. There was an ice-free corridor between this ice mass, and the Scandinavian ice mass to the north.

The Siberian ice sheet, on the other hand, was confluent with those of the Urals and Scandinavia, though it was smaller than the latter, and failed to reach as far south. The over-all extent of the glaciers decreased as one moved east, largely because of the absence of a suitable source of moisture and energy.

The Scandinavian ice sheet (Fig. 2.8), at its greatest known maximum, was probably coalescent with ice spreading from the Ural Mountains of Russia, and, in the south-west, with glaciers of British origin. It extended an unknown distance into the Atlantic off Norway, and may have merged with ice over Spitzbergen. In the south, the Elster and Saale glacial borders trend along the northern bases of the central European Highlands, and the Saale ice sheet, penetrated far down the basins of the Dnepr and Don Rivers.

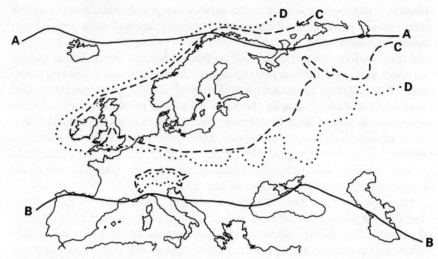

FIG. 2.8. Glacial conditions in Europe
  A. The position of the present polar timberline in Europe.
  B. The position of the timberline during the maximum of the last (Würm) glacial.
  C. The extent of north European drift deposits that can be attributed to the Last Glacial.
  D. Represents the drift borders of the Riss-Saale and Mindel-Elster in North Europe (from data in Flint, 1971 and Kaiser, 1969).

Although the thickness of this ice sheet is not known with precision, it may well have exceeded 3000 m both over the Sognefjord region of western Norway, and at the head of the Gulf of Bothnia, though the average thickness was probably about 1900 m.

The ice also extended across the present North Sea basin, which in times of maximum glaciation was largely above sea-level (see p. 197). The Dogger Bank, which rises some 20 m higher than the surrounding sea-floor, is possibly the remnant of a great moraine some 250 km long, and 100 km wide (Stride, 1959).

The southern continents

In the southern continents much less work has been done on the nature of the glacial periods, though it is clear that glaciation was markedly less important in its areal extent than it was in the northern hemisphere.

For example, the Drakensberg Mountains of South Africa seem to have been largely unaffected by glacial activity (though relict frost effects have been noted), and in Australia glacial activity, as a result of the relatively low relief and arid interior, was sharply restricted. Former glaciation of the mainland was confined to a single zone in the Snowy Mountains with an area of barely 52 km², though in Tasmania glaciation was more extensive, with an ice cap in the Central Plateau. New Zealand, with its greater relief and oceanic conditions, has some glaciers at the present day, unlike

Australia, and in the Pleistocene the New Zealand Alps in the South Island were intensively and widely glaciated, though the North Island was largely unaffected. In South America the ice sheets, developed from the great Andean Cordillera, were much larger, and in the far south the glaciated zone was over 200 km wide, and the ice may have been over 1200 m thick. A more or less continuous zone of glaciation extended to about 30 °N, though north of 38° the ice tended not to extend very far from the Cordillera, either to the Pacific in the west or into the plainlands of the east. The furthest north ice body was that which capped the Sierra Nevada de Santa Marta in Colombia.

The glaciation of the Antarctic is as yet very imperfectly known, though it is clear that both the margins and the thickness of the ice sheet varied during the course of the Pleistocene and Late Tertiary. A large ice sheet existed in western Antarctica as early as the Eocene (see p. 20), and signs of glacial action on nunataks [1] suggest that the ice may well have been 300–800 m thicker than at present, though the over-all lateral extent of this great ice mass was controlled to a considerable degree by calving of icebergs into the relatively deep waters offshore. Some expansion may have resulted from lower glacial sea-levels.

From time to time it has been proposed that the glaciers may have fluctuated out of phase with those of the northern Hemisphere. According to this postulate, worldwide interglacial conditions would allow relatively warm and moist air to penetrate Antarctica, thereby increasing by a considerable degree the rate of accumulation of snow, and the degree of expansion. Although the climatic fluctuations of the last 200 years lend some support to this view of non-contemporaneous glacial history in Antarctica, the so-called Ross I glaciation has been dated at about 9500 to 35 000 B.P., which ties in with the late Wisconsin glaciation (Flint, 1971, p. 713). Moreover, isotopic investigation of an ice-core from Byrd Station suggests a synchronism between major climatic events in Antarctica, and those in the northern Hemisphere, with a major cold phase culminating at about 17 000 yrs B.P. and terminating at about 11 000 B.P. (Fig. 2.13 A, B).

Similar dates for the end of the last glacial phase have recently been obtained by pollen-analysts in numerous parts of the southern hemisphere and in some equatorial areas, including Kenya, Colombia, Fuego-Patagonia, southern Chile, the Argentine basin, Marion Island, and New Zealand. These tend to substantiate the view of the contemporaneity of major events in the two hemispheres.

Permafrost and its extent in the Pleistocene

Beyond the limits of the great Pleistocene ice sheets there were, particularly in Europe, great areas of open tundra. These areas were frequently underlain

---

[1] A nunatak is an isolated hill or peak remaining above the level of an ice sheet.

by permafrost. Permafrost is a frozen condition in soil, alluvium, or rock, and is currently concentrated in high northern latitudes, reaching thicknesses of as much as 1000 m.

The current southern boundary of *continuous permafrost* coincides approximately with the −5 or −6 °C mean annual isotherm. The limits of *discontinuous* and *sporadic* permafrost are rather higher, but mean air temperatures have to be negative. Thus in Europe continuous permafrost is restricted to Novaya Zemlya, and the northern parts of Siberia, whilst discontinuous permafrost extends into northern Lappland. However, there is very strong evidence for the former extension of such permanently frozen subsoil conditions to wide areas of Europe during the glacial cold phases. The evidence consists of the casts of ice wedges which form polygonal patterns in areas of permafrost. The casts can either be identified in sections, or detected as crop marks on air photographs. These have been encountered very widely, for instance in southern and eastern England, especially in Kent and East Anglia, and they are even fairly extensive in the valley of the Severn and Warwickshire Avon, and in parts of lowland Devon. The only part of mainland Britain to have been unaffected by permafrost is probably the extreme tip of the south-western peninsula (Williams, 1975).

With regard to the former southern extension of permafrost in mainland Europe there is considerable dispute, though even the most northerly proposed boundary indicates that only the central and southern Balkans, peninsular Italy, the Iberian peninsula, and south-west France were largely unaffected. This indicates very forcibly the degree to which tundra and periglacial conditions were displaced southwards, and the extent to which temperature conditions were depressed in much of Europe. On the basis of a −5 °C limit for permafrost it seems likely that in eastern England, where conditions in the Pleistocene may have been made more continental by the drying out of the North Sea during glacial low stands of sea-level, mean annual temperatures were depressed by 15° (or more) during the Last Glaciation.

In North America relatively less is known about the southern displacement of the permafrost zone, but it seems likely that as the southern extent of the Wisconsin ice sheets was further south than that of the Würm–Weichselian in Europe, the zone of more severe periglacial conditions was probably more restricted.

Although the presence of permafrost suggests that mean annual temperatures were at least as low as those now experienced in tundra areas, it is probable that the periglacial climates of glacial Europe and America were different in character from any now found on Earth. Because of latitude, especially in America, days were longer in winter and shorter in summer than in any high latitude periglacial area at present. Also, the sun would have risen higher in the sky, giving both higher mid-day temperatures, and more marked diurnal changes. Evaporation rates would have been higher.

The formation of loess sheets

Around the great ice sheets, great spreads of loess were deposited during the course of the Quaternary. Loess is a largely non-stratified and non-consolidated silt, containing some clay, sand, and carbonate, which was deposited primarily by the wind (Smalley and Vita-Finzi, 1968). It is markedly finer than aeolian sand. Over vast areas (at least 1.6×10^6 km^2 in North America and 1.8×10^6 km^2 in Europe), it blankets pre-existing relief, and in China has been recorded as reaching a thickness of 180 m. In the Missouri Valley of Kansas the loess may be 30 m thick, in European Russia sustained thicknesses of 10 to 15 m are found, along the Rhine thicknesses approach 30 m, in Argentina thicknesses, often 10 to 30 m, reach over 100 m in places, while in New Zealand, on the plains of the South Island, thicknesses reach 18 m.

The sources of loess include desert basins, but exposed outwash and areas of till recently uncovered by deglaciation, are probably the most important. The winds, some of which may have blown away from the ice sheets with great velocity, moved the finer materials, and these were then deposited as loess at some distance, especially where there was a dense vegetation cover, as along river valleys, to trap it.

The distribution of loess is now well known, and the main areas in America include central Alaska, southern Idaho, eastern Washington, north-eastern Oregon, and even more important, a great belt from the Rocky Mountains across the Great Plains and the Central Lowland into western Pennsylvania. Loess is less prominent in the eastern U.S.A. as relief conditions for deflation, and the nature of outwash materials, seem to have been less favourable than in the Missouri–Mississippi region. In Europe the loess is most extensive in the east, where, as in the case of America, there were plains and steppe conditions. The German loess shows a very close association with outwash, and in France the same situation is observed along the Rhone and Garonne Rivers. These two rivers carried outwash from glaciers in the Alps and Pyrenees respectively. The Danube was another major source of silt for loess in eastern Europe. Britain has relatively little loess, and this may result from the oceanic climate which would tend to reduce the area of exposed outwash. Indeed, in Britain wind-lain sediments of peri-glacial age are conspicuous only for their rarity—dunes are low, rather shapeless hummocks occurring only in a few localities, cover sands are thin and patchy in comparison with those in the Netherlands, and 'loess is more of a contaminant of other deposits than one in its own right' (Williams, 1975). The maximum depth of loess in Britain is only about two or three metres, and sand dunes of periglacial origin are restricted to a few small areas, including the Breckland of East Anglia, and the Scunthorpe and York areas. Some sands, such as the interstadial Cheltenham Sands, may be banked up against escarpments.

In Asia, the steppes and deserts of the interior may have been the source of the great deposits in China. In South America, where the Pampas of Argentina and Uruguay has thick deposits, a combination of semi-arid and arid conditions in the Andes rain-shadow, combined with glacial outwash from those mountains, created near ideal conditions. However, in Australia and Africa, where glaciation was relatively slight, loess is much less well developed.

The importance of this loess for the settlement of post-glacial Europe is referred to on page 119.

The degree of climatic change in glacials and pluvials

Although the presence of greatly expanded ice sheets, and of permafrost conditions, gives a broad indication of the extent to which temperatures changed during the glacial intervals of the Pleistocene, it is possible, through a variety of techniques, to gain some more precise and quantitative measures of the degree of climatic change that has taken place.

Temperatures can be assessed through five main lines of evidence: isotopic measurements, the levels of cirques, the extent of permafrost, the limits of frost-affected sediments, and the nature of floral and faunal remains. These methods are all subject to certain difficulties and pitfalls in that temperature may be only one of the controls which influence, say, the position of the tree and snow lines. Similarly, the interpretation of the palaeoclimatic significance of snowline levels, represented by cirque floor heights, depends very much on the estimation of probable local lapse rates. Lapse rates are the mean rates at which temperatures change with altitude (generally $0.6\,°C/100$ m), but they are subject to local fluctuations.

The isotopic methods, which include an examination of O^{18}/O^{16} ratios of fossil Foraminifera, have proved fruitful, especially as they have been applied to deep-sea cores, though here too there are two main factors involved: the temperature of the ocean water, and the original isotopic composition of the ocean water. There has been much discussion as to the relative importance of these two factors (Shackleton, 1967). In principal, however, there is a relationship between the relative abundance of the two oxygen isotopes O^{16} and O^{18} in biogenic carbonates (mollusc shells, tests of Foraminifera), and the water temperature at the time that the carbonate was formed. O^{18} enrichment increases 0.02 per cent per $1\,°C$ temperature fall, and this small change in the ratios can be detected by mass-spectrometer.

The downward movement of snowlines in glacial ages indicates lowering of temperatures, especially summer temperatures, though it has to be remembered that precipitation and cloudiness could also affect the level, as does the local lapse rate. A knowledge of local lapse rates is required to relate the altitudinal shift of the snowline to temperature change. The position of the Pleistocene snowline is also subject to some error in its assessment, in that it is determined by a study of the position of cirque floors. Cirque floors tend to cluster at or just below the 0° summer isotherm. In general the cirque floor measurement is only valid in areas where the former glaciers never grew beyond the corrie type. Values that have been determined by this method suggest a mean temperature depression during glacial phases of the order of 5 °C. The varying degree of snowline depression from region to region gives a spread in temperature depression values of from only about 2·0 °C to over 10 °C. This reflects the fact that snowline depression values ranged from as little as 600–700 m in the northern Urals, the Middle Atlas, and the Caucasus, to as much as 1300 to 1500 m in the northern Pyrenees, on Kilimanjaro, in the Apennines, and in the Tell Atlas.

The former extent of permafrost has been discussed in relation to Europe on p. 42. In that the current boundary of permafrost in Siberia, Scandinavia, and North America can be related to mean annual temperatures, it is possible to infer Pleistocene temperatures from this source. The permafrost data tend to give somewhat higher values for the amount of temperature depression than do the snowline data, with a value of 15 or 16 °C being recorded for the Midlands of England and for East Anglia, values of 10–15 °C for parts of central North America, and a value of 11 °C for Germany.

In many parts of the world, where conditions are now both too warm and too dry for frost activity to be important in rock disintegration, there are screes of angular debris, which have been widely interpreted as being the product of frost activity. These have for example, been described from Cyrenaica and Tripolitania in Libya (Hey, 1963). Periglacial deposits of this type have been used to suggest a 11 °C depression of glacial temperatures in the south western U.S.A. (Galloway, 1970), a greater than 9 °C depression in the Snowy Mountains and Canberra regions of Australia, and a depression of over 10 °C in the Cape Province of South Africa.

The data provided by organisms and plants are extremely difficult to interpret other than qualitatively, though from various sources Flint has suggested that at the height of the Last Glacial, temperatures may, on average, have been depressed by about 6 °C, a value which ties in with the snowline evidence, though depressions of 10–15 °C have been suggested for Central Europe (Segota, 1966).

Foraminifera from the glacial segments of deep-sea cores, when compared with those of the present in the same locations indicate a temperature depression in glacials of about 5 °C for Caribbean surface-water, 4·6 °C for the Equatorial Atlantic, and 5·7 °C for equatorial waters off West Africa (Hecht, 1974). A recent general review of the sea-water temperature evidence (Climap Project Members, 1976) has suggested that on a world basis the average anomaly between present and glacial surface-water temperatures was of the order of 2·3 °C. However, locally, as in the North Atlantic, where the position of the Gulf Stream appears to have shifted substantially, the values may have been from 12 to 18 °C different from today.

On land as well, local temperature depressions may have been greater than has been suggested hitherto. Areas subjected to an ice covering, because of the temperature gradients, and highly reflective conditions associated with ice caps, may have become as cold as Antarctica, probably cooling by as much as 60 °C to an average temperature as low as − 60 °C.

The calculation of former precipitation levels is even more beset with difficulties than the calculation of temperature changes, in that most of the methods attempt in fact to measure not precipitation, but evaporation/precipitation ratios. They are thus partially dependent on temperature estimation. Decreases in temperature of the type discussed above would in many areas be sufficient by themselves to account for certain 'pluvial' or lacustral phenomena which have been interpreted in the past as being the result of increased precipitation. Phenomena that can be used to assess changed precipitation/evaporation ratios include the volumes and stratigraphy of lakes (see p. 73), the nature of cave fillings, the distribution of dune fields (see p. 67), the characteristics of fossil soils (palaeosols), and the nature of former stream regimens as deduced from sedimentological and morphological evidence. It is not easy to obtain any quantitative data from these sources though various attempts have been made.

As noted on p. 6 many lakes had higher volumes at some stage in the Pleistocene and Early Holocene. In a closed basin the level of the lake depends on the balance between rainfall inputs, evaporation, and surface area. One of the main controls of evaporation rates is temperature, so that if this can be estimated for the Pleistocene it is possible to calculate how much rainfall is needed to account for calculated volumes and areas of lakes at various levels above their current ones. On this basis, for example, it has been calculated that rainfall totals must have been about 165 per cent of totals of the present day in East Africa from 9000–6000 B.P., assuming that temperatures were 2–3 °C lower during Early Holocene times than at present (Butzer *et al.*, 1972).

In America, using snowline-derived temperature data and related measures, several geologists and hydrologists have recently appraised the water budgets of various pluvial lakes in the Basin and Range Province during their maximum, Late Pleistocene levels. Estimates of the increase in mean annual precipitation (from present average values over the drainage areas, compared with those during the lake maximum) ranged from 180–230 mm, and for the decrease in mean annual temperature, from 2·7 to 5 °C. Thus, for example, at Spring Valley in Nevada, Snyder and Langbein (1962) proposed a pluvial rainfall of 510 mm compared with about 300 mm at the present. It needs, however, to be stressed that these estimates are based on low temperature depression values. By contrast, on the basis of temperatures implied from periglacial features, Galloway (1970) has proposed that in the south-west United States temperatures were depressed by 11 °C. He calculated on this premise that far from precipitation levels being higher during the pluvial phases, precipitation levels were only about eighty to ninety per cent of current levels. On the other hand a general survey of the evidence, notably in Australia and the United States, has led Dury (1967) to propose that shrunken lake levels and misfit streams (streams too small for their valleys) indicate an increase in mean annual precipitation during pluvials of the order of 1·5 to 2 over present totals; even when allowance is made for temperature reductions.

The vegetational conditions of the full glacials in Europe

During the various, full glacial stages of the Pleistocene, the vegetation of much of unglaciated, periglacial Europe, was characterized by its open nature. Trees were relatively rare, and in many respects the plant assemblages displayed many characteristics one would expect in a cool 'steppe' environment (Fig. 2.9)

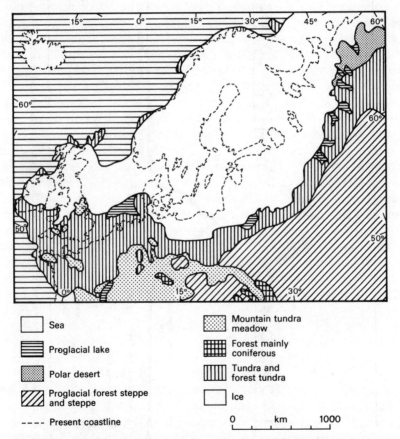

FIG. 2.9. Palaeogeographic reconstruction of northern Europe during the maximum of the Valdai (Last Glacial) (after Gerasimov, 1969, Fig. 3).

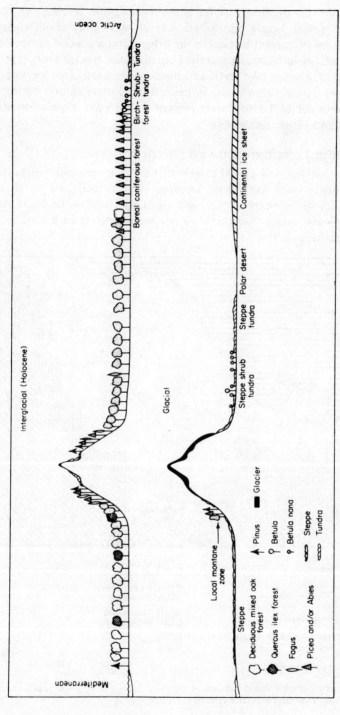

FIG. 2.10. Schematic representation of the vegetation during an interglacial and during a glacial, in a south–north section through Europe (from Hammen et al., 1971, Fig. 6).

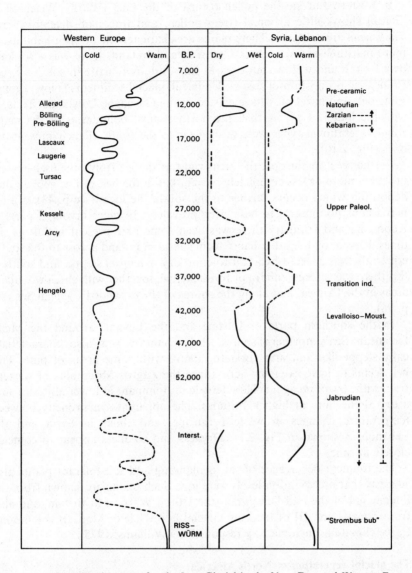

FIG. 2.11. Palaeoclimatic curves for the Last Glacial in the Near East and Western Europe, based on pollen-analysis (from Leroi-Gourhan, 1974, Fig. 1). Some interstadial names are given for Western Europe. Arcy, Tursac, Laugerie, and Lascaux are also names of prehistoric cave sites. In the right hand column certain names of cultural phases are given.

In western Europe the pollen record of the Last (Würm, Weichselian) Glacial shows little arboreal (tree) pollen, and traces of *Artemisia*[1] and *Thalictrum* are common. These plants are characteristic of open habitats. In more maritime areas, such as Cornwall and Ireland, there was also some dwarf birch and willow, but even as far south as Biarritz in south-western France the proportion of tree-pollen in full glacial sediments is low, though some oak and hazel may have existed in the Gascogne Lowlands. Thus, as with the limits of permafrost, the northernmost boundaries of the major zonal vegetational types were pushed far to the south of their present-day ones (Fig. 2.10).

Further east in Europe, the areas right at the ice fronts themselves were probably more or less completely barren, but in the belt of fine aeolian dust deposition which occurs further to the south, the loess (see p. 43), a more herbaceous flora seems to have been prevalent. In more favoured parts of Roumania and Hungary there was even some pine present in full glacial times. Russia, on the other hand, from southern Poland across to the southern Urals, was covered by a salt-tolerant dry *Artemisia* steppe, and south of this there was a forest tundra or forest steppe, together with small woodland areas in the Crimea, and along the shores of the expanded Caspian Sea (see p. 76).

In the southern parts of Europe and the Levant, around the Mediterranean Sea's northern shores, the vegetation was also characteristically steppe-like and arid (Bonatti, 1966), with some areas of pine. This belt seems to have extended across into the Zagros Mountains of western Iran, with *Artemisia* again characteristic or dominant at lower altitudes, and a dry alpine flora at higher altitudes. The apparent synchroneity between temperature changes in western Europe, and those in Syria and the Lebanon, is shown in Fig. 2.11. Dryness and coldness appear to coincide closely in time.

The frequent occurrence of salt-demanding and salt-tolerant plants also suggests that precipitation levels were low. Such plants are known from the interstadial of the mid-Devensian (the Upton Warren) in Britain, and also from Zones I and III of the Late Glacial in the Isle of Man. 'It is tempting to see climate as in some way responsible' (Williams, 1975).

The glacial vegetation of North America

While much of the country north of the European Alps during the Last Glacial maximum supported tundra or, close to the ice, a cold rock desert, in America the available records indicate that much of the area south of the ice sheet was covered with boreal forest rather than with tundra. The reason for this difference is that the ice limit in Wisconsin times in America was much further south than it was in Europe—39 °C in Illinois compared with

[1] A common name for *Artemisia* is worm-wood.

52 °N in Germany. Further, the Alps, with their own large ice cap, reinforced the semipermanent area of high atmospheric pressure associated with the Scandinavian ice sheet. This probably tended to divert warm, westerly air flows to the south of the Alps. There is no such mountain mass trending east to west in North America.

The widespread boreal forest of full glacial times in North America, dominated generally by *Pinus* and *Picea*, was not found everywhere, for there were patches of tundra and of treelessness, but these were not as extensive as in Europe. The southern limit of the boreal forest is not known with any certainty, but it was probably somewhere in south-central United States, perhaps extending westward from Georgia. Thus it may have formed a latitudinal belt as broad as it is today—1000 km from Hudson Bay to the Great Lakes. In the south-west, where pluvial lakes appear to have been synchronous with the main glacial (see p. 74), pollen evidence indicates high percentages of *Pinus* and other montane conifers during the Wisconsin, whereas now the same areas are characterized by semi-desert shrubs. In the Western Cordillera the tree line was lowered 800–1000 m, and the extent of alpine vegetation in the mountains was very greatly expanded.

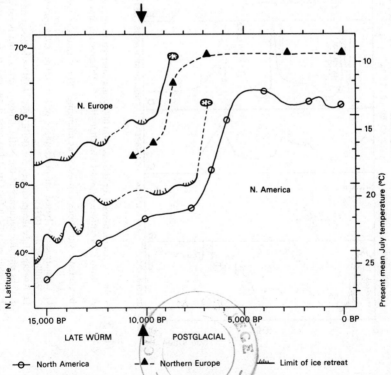

FIG. 2.12. Latitudinal changes of the arctic timberline since 15 000 B.P. (after Markgraf, 1974, Fig. 4).

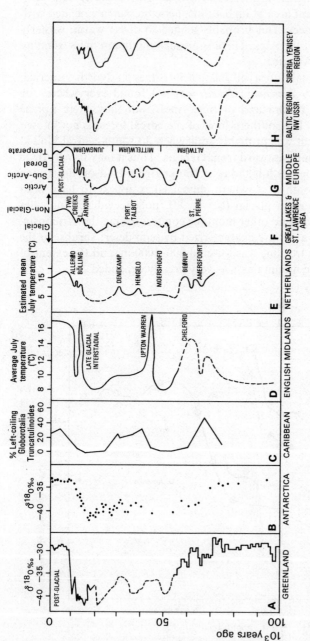

FIG. 2.13. Climatic fluctuations over the last 100 000 years inferred from miscellaneous evidence.

A. Climatic variations expressed as changing $^{18}O/^{16}O$ ratios in an ice core from Camp Century, Greenland (after Dangsgaard *et al.*, 1969).

B. Climatic variations expressed as changing $^{18}O/^{16}O$ ratios in the Byrd Station Ice Core, Antarctica (after Epstein *et al.*, 1970).

C. Climate curve based on percentages of left-coiling *Globorotalia truncatulinoides* in a Caribbean core (A 179–4) (after Wollin *et al.*, 1971).

D. Average July temperatures for the English Midlands based on the study of beetle faunas (after Coope, 1975).

E. The climatic sequence in the Netherlands inferred from floral evidence (after Hammen *et al.*, 1967).

F. Glacial activity in the Great Lakes/St. Lawrence region according to Dreimanis and others (after Fling, 1971).

G. Glacial and other fluctuations in Middle Europe (after Mörner, 1969).

H. and I. Glacial fluctuations in the U.S.S.R. (after Dreimanis and Raukas, 1975).

In all the diagrams colder conditions are indicated by a move of the curves to the left.

Likewise the latitudinal position of the Arctic timberline in the Late Würm (see Fig. 2.12) was very different from that of post-glacial times (Markgraf, 1974), the shift being about 24–25°.

The interstadials of the Last Glacial

One of the problems of glacial correlation is that of phases of lesser glaciation and relatively greater warmth during the course of a major glacial phase. Such interruptions are called interstadials, but there is as yet no universally acceptable definition which differentiates an interstadial from an interglacial. Nevertheless, there are indications in many parts of Europe, and elsewhere, that the Würm–Weichsel–Wisconsin glaciation was interrupted by certain phases of less intense glacial activity (Fig. 2.13), which enabled soils and other distinctive sediments to develop. Quite a large number of these deposits have now been dated by radiocarbon means, and certain correlations seem possible (Fig. 2.14). Examination of these dates suggests that while there is a considerable spread in values, there is some clustering over a period from about 50 000 to 23 000 B.P. This period was probably not a continuous phase of relatively warmer conditions, and there seems in many areas to have been a tendency for a particularly marked interstadial at the end of this time, notably around 28 000 B.P. (Denekamp, Plum Point, Paudorf, Kargy, Olympia, etc.). There were also some relatively short-lived interstadials near the beginning of the Würm–Wisconsin–Weichsel, and these may have been sufficient to lead to deglaciation in Scandinavia (Brörup, Amersfoort, Chelford, St. Pierre).

The period from 25 000 B.P. until the end of the Pleistocene saw a great expansion of glaciers, at least in the northern hemisphere, and this has been given a series of local names, including *Hauptwürm* in Europe, Woodfordian in the northern United States, and Pinedale in the Rockies. The last few millennia of the Last Glacial were also marked by some minor stadials and interstadials, and these are described on p. 95.

There is also confirming evidence from ice cores, both from the Arctic and Antarctic, of interstadials (Fig. 2.12). At Camp Century, Greenland, warmer periods at 19 000–23 000, 46 000–56 000, 68 000–74 000, and 75 000–78 000 B.P. have been suggested, whilst at Byrd Station in the Antarctic there are signs of warmer phases at 25 000, 31 000, and 39 000 with colder phases culminating at 27 000, 34 000, and 46 000 yrs. B.P. In Atlantic and Caribbean deep-sea cores there were maxima of warm conditions around 25 000, 40 000, and 65 000 years ago.

Pollen and faunal evidence has been utilized for assessing environmental conditions during the interstadials. In England, for example, the Chelford interstadial of around 60 000 years ago was characterized by boreal forest, and a beetle fauna comparable to that found in south-east Finland at the present day. The fauna of the Upton Warren Interstadial of around 40 000 years ago suggests July temperatures at least 5 °C higher than those which

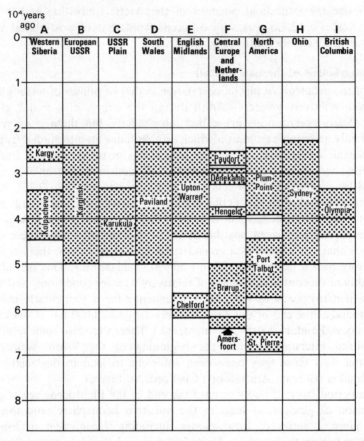

FIG. 2.14. Radiocarbon dated interstadials of the Last Glacial period in the northern hemisphere. Interstadials of the Late Glacial are not included. Drawn from data in
Zubakov, 1969. (A), Kind, 1972. (B), Serebryanny, 1969. (C), Bowen, 1970. (D),
Coope, 1975. (E), Segota, 1967, Butzer, 1964, Hammen *et al.*, 1967. (F), Dansgaard
et al., 1971. (G), Dreimanis *et al.*, 1966. (H), Fulton, 1968 (I).

existed during the full glacial conditions which followed. Indeed, in a recent
review of the evidence provided by coleoptera, Coope (1975) has suggested
that at the thermal maximum of the Upton Warren Interstadial, perhaps at
around 43 000 B.P., average July temperatures in central England were
about 18 °C, which is a little warmer than now. He does, however, find that
winter temperatures were somewhat lower than now, indicating a more
continental climatic regime. The warm phase was relatively short-lived, possibly lasting only a thousand or so years. In Denmark the early Weichselian
Brörup Interstadial had July temperatures only 2·3 °C lower than those of
the present day, and in the Netherlands it has been suggested that temperatures were much the same as those of the present.

The effects of the relative warming of the various interstadial phases can be seen in the vegetation of Europe at those times, as determined by pollen-analysis. The steppe belt of southern Europe, dominated in the full glacials by *Artemisia*, showed higher arboreal pollen characteristics. During the Denekamp interstadial there were pine forests in both southern Spain and in Macedonia, whilst in the Brörup interstadials a *Quercus ilex* vegetation was present in southern Spain, but a *Carpinus-Ulmus-Tilia* forest was present in Macedonia. In general woodlands returned to quite wide areas of Europe, with boreal woodlands of spruce, pine, and larch bordering the North Sea and Baltic. Oak and hornbeam forests probably occurred in northern Italy, Yugoslavia, and Albania.

The nature of the interglacials

In general terms the interglacials, to judge from the results of pollen-analysis and other techniques, appear to have been essentially similar in their climate, flora, fauna, and landforms to the Holocene in which we live today. The interglacials appear to have varied in length, and Butzer (1975) suggests that 'warm' stages over the past 800 000 years have had a duration of from 23 000 to 73 000 years. On the other hand, the United States Committee for the Global Atmospheric Research Program (1975, pp. 178–9) have a different point of view. 'Few climatologists', they say, 'would dispute that the prominent warm periods (or interglacials) that have followed each of the terminations of the major glaciations have had durations of 10 000 ± 2000 years.' Whatever their length may have been, and this varies according to whether one holds the view of a 'long' or a 'short' Pleistocene, the most important of the characteristics of the interglacials was that they witnessed the retreat and decay of the great ice sheets, and saw the replacement of tundra conditions by forest over the now temperate lands of the northern hemisphere (Fig. 2.15). Tree-lines occurred at higher altitudes and latitudes (Fig. 2.12).

The maximum temperatures attained in some or most of the interglacials appear to have been a little higher than those of the present, and may well have been comparable to those of the Holocene climatic optimum (see p. 114). During the Last or Sangamon Interglacial, for instance, much of North America was covered by deciduous forest as at the present, though near Toronto in Canada the presence of pollen from the sweet gum (*Liquidambar*)[1] suggests that temperatures may have been 2 to 3 °C higher than those now experienced in that area. Equally, in the previous Holstein Interglacial of Poland and Russia, the fauna, particularly the distribution of beech, hornbeam, holly, and the Pontic alpine rose, suggest temperatures slightly warmer than those of today. During the Hötting Interglacial the flora indicates a temperature about 3 °C higher than that of the present, and

---

[1] Table 2.5 is a list of common plant names, with their botanical equivalents, as used in this section.

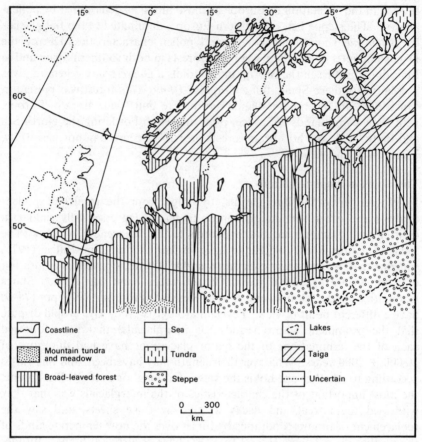

FIG. 2.15. Palaeogeographic reconstruction of northern Europe during the Last (Mikulino or Eem) Interglacial (after Gerasimov, 1969, Fig. 2).

subtropical woodland and forest seem to have been more extensive in Italy, the Balkans, and the Caucasus, implying that conditions may have been rather moister as well.

The general sequence of vegetational development during the interglacials has been rationalized by Turner and West (1968), who propose the following type of pattern as being characteristic:

(a) The first phase, one of climatic amelioration from full glacial conditions, can be called the Pre-temperate zone. It is characterized by the development and closing in of forest vegetation, with boreal types being dominant. *Betula* and *Pinus* are a feature of the woodlands, but light-demanding herbs and shrubs are also a significant element of the vegetation. Relicts of the preceding Late Glacial periods such as *Juniperus* and *Salix* may also be present.

(b) The next phase, termed the Early-temperate zone, sees the establishment and expansion of a mixed oak forest with many shade-giving forest genera, typically *Quercus*, *Ulmus*, *Fraxinus*, and *Corylus*. Soil conditions were generally probably good, with a mull[1] condition, and this promoted dense, luxuriant cover.

(c) In the next phase, the Late-temperate zone, there is a tendency for the expansion of late-immigrating temperate trees, especially *Carpinus*, *Abies* and, sometimes, *Picea*, accompanied by a progressive decline of the mixed oak forest dominants. Some of these changes may be related to a decline in soil conditions, associated with the development of a mor[2] rather than a mull situation.

(d) The fourth phase is called the Post-temperate phase, and is indicative of climatic deterioration. There is a reduction of thermophilous genera, and an expansion of heathland. The forest becomes thinner, with temperate forest trees becoming virtually extinct, and a return to dominance of boreal trees, such as *Pinus*, *Betula*, and *Picea*.

This general sequence, whilst broadly applicable to the main interglacial phases, does vary from interglacial to interglacial. There were probably climatic differences between the various phases, different barriers to migration, differing distances of glacial refuges from which genera expanded, changes in ecological tolerance, and variability within genera, and other changes consequent upon evolution or extinction (West, 1972, pp. 315–24).

One very pertinent question to raise about vegetation successions in interglacial times is to ask at what speed the colonization of trees was able to progress across country. From Sweden it would appear that in the Holocene Interglacial Scots pine and pubescent birch spread at a rate of 205–60 m per year, alder at 175–230 m, elm and warty birch at around 190 m, and hazel at 130 to 190 m per year. As a whole it would seem that a rate of advance of some 200 m per year for trees with light seeds, and, rather less, say 160 m, for trees with heavier seeds, like hazel and oak, was characteristic. Thus advance of trees at the end of Late Glacial times could have taken place at about one kilometre in five years, or 1000 km in 5000 years.

Plainly, there would be very different response curves with time for climate, glaciers, and vegetation (Fig. 2.16). Ice caps would tend to respond relatively sluggishly to a climatic amelioration because of their great mass, and because of their partial control of regional climates. Retreat rates of up to 3 km/100 years for the Greenland Ice Cap are much lower than the rates of floral advance discussed above. The response of fauna would tend to be even quicker.

[1] *Mull* is fertile, non-acidic soil humus inhabited by earthworms.
[2] *Mor* is an acid soil humus which accumulates at the soil surface and is too acid for earthworms.

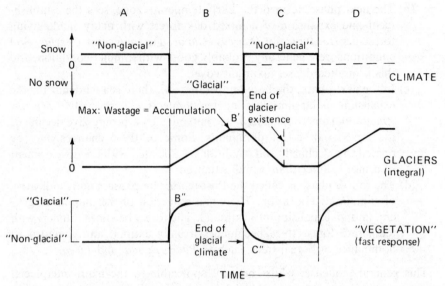

FIG. 2.16. The different response curves, with time of climate, glaciers, and vegetation (after Bryson and Wendland, in Andrews, 1975, Fig. 5–6).

Variations in the British and European interglacials

The interglacial temperate forests recorded in the early British interglacials, the Ludhamian and Antian (see Table 2.4 and Fig. 2.3 for the location of these phases in the local sequence) are mixed coniferous and deciduous, and the presence of hemlock and wingnut makes the vegetation assemblage different from that of any subsequent period in Britain. The relatively severe Baventian Glacial period led to the extinction from the British flora of hemlock (*Tsuga*), though it has remained in the vegetation associations of parts of North America until the present day. In northern Eurasia certain of the late Pliocene vegetation types do not appear to have reoccupied the area after the first severe cold of the Pleistocene, though they were present in the Tiglian (Ludhamian). These plants included *Sequoia*, *Taxodium*, *Glyptostrobus*, *Nyssa*, *Liquidambar*, *Fagus*, *Liriodendron*, and several others. Thus a marked degree of impoverishment took place in the British flora as a result of the oncoming of the first cold phases of the Pleistocene (West, 1972).

With regard to the later British interglacial floras, however, the Cromer Forest Bed of the Cromerian Interglacial on the Norfolk coast resembles the contemporary British flora much more closely, and Yew (*Taxus*), *Quercus*, Beech (*Fagus*), *Carpinus*, *Ulmus*, *Betula*, and Hazel (*Corylus*) can be identified in the bed. The Hoxnian Interglacial displays in its pollen-profiles a high frequency of *Hippophaë* at the beginning of the sequence, then a late rise of *Ulmus* and *Corylus*, and the presence of *Abies* and *Azolla filiculoides*.

Table 2.6
Botanical names of interglacial plants

Botanical name	Common name
Abies	Fir
Acer	Maple
Alnus	Alder
Azolla	Water fern
Buxus	Box
Carpinus	Hornbeam
Corylus	Hazel
Erica	Heath
Fagus	Beech
Fraxinus	Ash
Juglans	Walnut
Juniperus	Juniper
Lemna minor	Duckweed
Liquidambar	Sweet gum
Liriodendron	Tulip tree
Osmunda claytonia	Fern
Picea	Spruce
Pinus	Pine
Pterocarya	Wingnut
Quercus	Oak
Salix	Willow
Sequoia	Sequoia
Taxus	Yew
Tilia	Lime
Trapa natans	Water chestnut
Tsuga	Hemlock
Ulmus	Elm
Vitis	Vine creeper
Xanthium	Cocklebur

Irish materials of the same age have a high percentage of evergreens, indicative of a high degree of oceanicity of climate (e.g. *Picea, Abies, Taxus, Rhododendron, Ilex*, and *Buxus*). The materials also contain certain Iberian species such as *Erica scoparia*, St. Dabeoc's heath (*Daboecia cantabrica*), and Mackay's heath (*Erica mackaiana*). The last two currently have a rather limited distribution in the Cantabrian mountains. The subsequent Ipswichian, on the other hand, seems to indicate rather more continental conditions. Its characteristic properties included an abundance of *Corylus* (hazel) pollen in its early part, and then much *Acer* pollen; also present was *Carpinus*, but there was a scarcity of *Tilia* in the second part of the Interglacial. Most of the Ipswichian sites contain a number of species not now native, including *Acer monspessulanum, Lemna minor, Najas minor, Pyracantha coccinea, Trapa natans, Xanthium*, and *Salvinia natans*. This suggests that conditions were somewhat warmer than during the Holocene climatic optimum (West, 1972, p. 310).

In the more continental parts of Europe the characteristic interglacial assemblages were a little modified, though the general sequence is not dis-similar. In the Likhvin (Holstein) of the U.S.S.R., for example, as in Turner and West's model, the first stage is represented by a high incidence of *Betula* and *Pinus* pollen, with *Picea* and *Salix* pollen only making up 1 per cent or less of the total (Ananova, 1967). This birch and pine–birch forest stage was replaced by a dominantly pine and spruce–pine sequence, with *Pinus* and *Picea* predominant, with *Betula* making up about 5–10 per cent of the total pollen, and *Alnus* pollen constantly occurring with a frequency of about 15 per cent. This coniferous forest was replaced by a combination of con-iferous, broad-leaved, and alder-thicket forests, with *Picea excelsa* domin-ant in the east, and *Pinus sylvestris* in the west. Both these types had about 40–60 per cent of the total pollen, with *Alnus* making up 25–40 per cent, and Quercetum mixtum (mixed oak) pollen about 10 per cent. *Abies* pollen was rare. In the next stage, *Abies* sometimes reaches 30 per cent, and *Carpinus* reached 20–30 per cent in some sections, though coniferous pollen was still dominant. Subsequently, there was a change back towards a more boreal flora, and this in turn was succeeded at the end of the Interglacial by a reduction in tree-pollen, the constant presence of various open vegetation plants (Poaceae, Cyperaceae, and *Artemisia*), and the arrival of plants of the periglacial type. Thus, although the sequence is comparable to that in western Europe, there are certain elements missing in north and eastern Europe, including *Vitis*, *Pterocarya*, *Juglans regia*, *Abies alba*, *Carpinus orientalis*, *Buxus*, *Taxus*, *Tilia tormentosa*, and *Osmunda claytoniana*.

An over-all picture of Europe during the Last Interglacial can be obtained from Fig. 2.15. This shows not only the vegetational characteris-tics of the Interglacial, including the great expansion of broad-leaved forest, but also the way in which the configuration of the Continent and of the Baltic Sea region was modified by the worldwide rise in sea-level occasioned by the melting of the great ice caps. By contrast, Fig. 2.9 illustrates the nature of Europe during the Last Glacial.

In southern Europe the interglacials may have been associated with moister conditions, in contrast to the glacials which were essentially drier. Palynological studies in southern Spain, for example (Florschütz *et al.*, 1974), show that instead of a steppe-like vegetation such as characterized the glacial periods, the interglacials were characterized by a more humid as-semblage of vegetation including *Fagus*, *Juglans*, *Quercus pubescens*, *Tsuga*, and *Cedrus*.

Faunal and floral fluctuations

The environmental changes of the Pleistocene, as already mentioned with regard to the changing nature of European vegetation in the various inter-glacials, led to a great impoverishment of flora, particularly of glaciated islands. It has been remarked, for example by Pennington (1969, p. 1), that the comparative poverty of the British flora, compared with that of continental Europe in comparable latitudes, is the result of successive wiping-out of frost-sensitive species by the repeated glacial episodes of the last million years. After each glacial period, with its wholesale extinction of plants from Britain, migrating plants and animals, followed northwards in the footsteps of the retreating ice, and combined with the descendants of the hardy species which had survived, to re-establish the British flora and fauna.

A slightly more complex situation is illustrated by the Irish fauna, where both glacial and sea-level fluctuations seem to have been important in determining the present types of animal encountered on the island. Ireland today does not possess certain beasts which are encountered in England and Wales. These include the poisonous adder, the mole, the common shrew, the weasel, the dormouse, the brown hare, the yellow-necked field mouse, the English meadow mouse, and others. On the other hand, it does possess a certain quite large proportion of the English fauna. The explanation for this seems to be that as the ice caps retreated animals from the Continent (then connected by dry land to England because of the low eustatic level at that time, see p. 197) and from the unglaciated tundra area of southern England, crossed over to Ireland by a landbridge. By the Boreal phase of post-glacial time, however (9500 B.P. onwards) when the climate had so ameliorated as to permit immigration of temperate species, the Irish Sea was in existence, and the dry passage to Ireland was disrupted. Thus many beasts were unable to cross.

A similar example of the role of the various Post-Glacial and Late Glacial events in creating the present pattern of fauna is provided by the distribution of bird species in the North American continent (Mengel, 1970). During the late Wisconsin glaciation, which reached its peak about 18 000 or 20 000 years ago, the northern Rocky Mountains were covered by the Cordilleran ice, while to the east, the lower ground was covered by the massive Laurentide ice sheet (see p. 37 for a further discussion of the extent of the American ice bodies). The two coalesced along the foot of the Rockies in Alberta, British Columbia, and the Yukon. There is evidence that as these two sheets contracted in Late Glacial times a long arm of tundra and then taiga invaded the lower ground from southern Alberta to the Mackenzie River Delta. The NW. to SE. orientation of this corridor helps to explain what is a peculiar but recurrent feature in the distribution of north American birds and some other animals, namely the strong tendency for essentially eastern taxa, that had adapted to the taiga and its successional stages, to occur north-west to, or nearly to, Alaska, at the apparent expense of western montane kinds that had adapted to montane coniferous forest.

The explanation seems to be that the western types were blocked by the persistent but dwindling Cordilleran ice sheet, enabling the eastern taxa to get there first, and to fill the niches: a situation which they have held since.

The question as to what degree the present fauna was able to survive in areas that were glaciated is one of great interest. On the one hand some authorities maintain that the bulk of the fauna of, say, Iceland, is of post-glacial age, and has reached that island by post-glacial diffusion. On the other, there are authorities who consider that certain species were able to exist on small non-glaciated peaks (nunataks), rising above the general level of the ice caps (Gjaeveroll, 1963). Other people consider that in certain favoured coastal regions there were small 'refugia' where a hardy flora might be able to live through the glacial period. The last two concepts comprise the *Overvintring* concept of certain Scandinavian botanists. That such survival is possible is attested by the flora of present-day Greenland nunataks. Moreover, various Scandinavian and Irish geomorphologists have claimed to find evidence for refugia and nunataks. One line of evidence that has been used in Arctic Norway is the presence of block fields (*felsenmeer*) and other periglacial rather than glacial features on summits. Moreover, notably in Iceland, the present distribution of the flora often shows a bicentric or polycentric form, which ties in better with the idea of diffusion from internal refugia than with the idea that the whole flora was erased— the *tabula rasa* concept—and has been replaced by post-glacial migrations from overseas. If post-glacial migration were responsible for these plants arriving one might expect them to be more widely distributed. Certain nunataks have also been proposed for Britain, including parts of the Cleveland Hills and the Pennine Chain.

The role of landbridges in the Pleistocene should not be exaggerated, though as will be seen (see p. 172), the fall in relative sea-level by perhaps as much as 150 m did expose large expanses of the continental shelf. Certain islands were therefore linked together, or to the mainland. Malta and Sicily, Capri and Italy, the Balearics, the Ionian Islands, and, possibly, Tunisia and Italy, are such examples. Other islands, on the other hand, remained isolated, and their fauna tends to this day to show a greater degree of endemism. This is well illustrated by an example from the Philippines. The Islands of Negros, Panay, and Masbate collectively make up Visaya. They stand together on a submerged shelf less than 50 m deep. They are, however, separated from the nearby island of Cebu by a 98 m deep strait. The main faunal result of this is that on Visaya there are 32 endemic species of non-migratory birds. They are lacking on Cebu. It thus seems likely that in this area the sea-level was low enough to permit migration among the three islands of Visaya, but not great enough to permit immigration to or from Cebu (Deevey, 1949).

Elsewhere in South-East Asia the effects of marine regressions and transgressions on faunal boundaries are also striking. Low stands of glacial sea-level drained most of the Sunda Platform area, consisting of Malaysia, Borneo, Java, and Sumatra. Thus these islands and peninsulas were interconnected and more than 3 million km² of shallow warm seas were

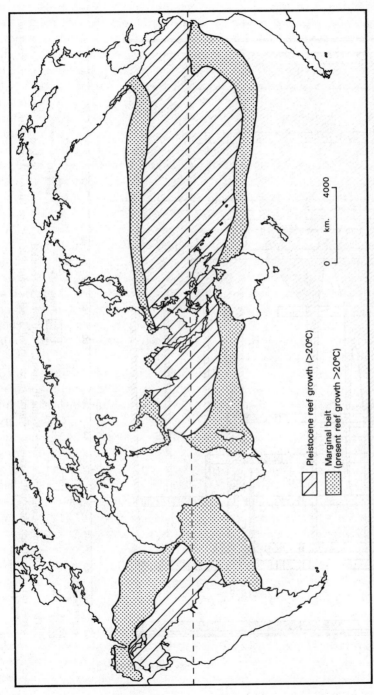

FIG. 2.17. Possible extent of the contraction of the coral reef seas during the Pleistocene. The isotherm of 20 °C is taken as the effective limit of reef formation, and the map is constructed by subtracting the glacial falls in temperature for each major ocean derived from published paleotemperature analysis from the present-day sea surface temperatures of the coldest month (from Stoddart, 1973 Fig. 3).

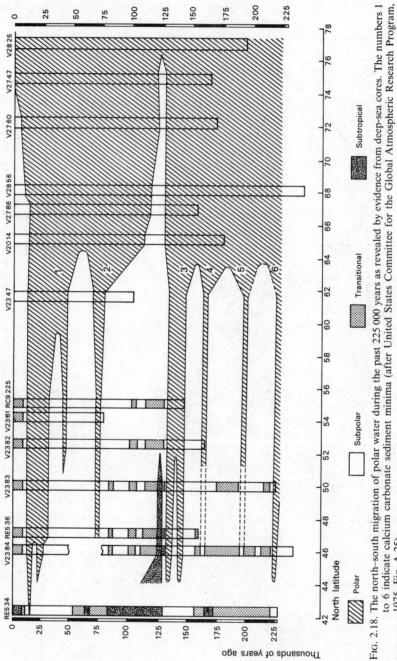

FIG. 2.18. The north–south migration of polar water during the past 225 000 years as revealed by evidence from deep-sea cores. The numbers 1 to 6 indicate calcium carbonate sediment minima (after United States Committee for the Global Atmospheric Research Program, 1975, Fig. A.25).

converted into land (Verstappen, 1975). This allowed many animals to come from the Asian mainland, including beasts of Indian and Chinese type. Today the fauna of the Sunda Islands is basically a somewhat impoverished version of that on the Asian mainland, with local races of elephant, tiger, leopard, and dhole. A low stand of sea-level may also have allowed some early humans, *Homo erectus*, to penetrate Java as much as one million years ago. However, the Sunda Shelf is separated from the New Guinea–Australia Sahul Platform by deep water, albeit narrow. Pleistocene sea-levels appear not to have fallen sufficiently to allow the linking of these two realms, and so this limited belt of sea forms one of the most important of all zoo-geographic boundaries, 'Wallace's Line'.

Unlike the Sunda Platform, the fauna of the Sahul Platform is one of distinctly Australian affinity, with marsupials—kangaroos, wallabies, wombats, and koalas. In between these two platforms is the group of islands called the Celebes. These appear to have been isolated from the two platforms for a considerable period, and so many types of animal are absent. Many indigenous forms such as the babirussa pig, the pygmy buffalo, and, in the fossil record, two types of pygmy elephant, have evolved.

The fall in temperature of the oceans, which was probably of the order of 3–8 °C during the cold phases of the Pleistocene also affected the distribution of marine life. This can be illustrated from a study of coral reefs (see Fig. 2.17). The present effective limit of reef growth is approximately that of the 20 °C ocean water isotherm (Stoddart, 1973). By subtracting the glacial falls in temperature for each major ocean derived from miscellaneous palaeotemperature observations a map of probable Pleistocene coral reef growth can be constructed. It shows the considerable degree of contraction which must have taken place in reef distribution. Over large areas reef corals would have died because of the relatively cool conditions. This effect would have been heightened further by the low still-stands of sea-level during glacial phases.

Similarly, the north–south migration of polar waters in the North Atlantic in response to major cycles of glaciation is shown in Figure 2.18. In this figure fourteen deep-sea cores have been arranged in a transect in the eastern North Atlantic. The boundary between the polar fossil assemblages (diagonal ruling) and the subpolar assemblages (open pattern) reflects the position of the oceanic polar front. At the glacial maximum about 18 000 years ago this front was some 20° farther south, while during the interglacial maximum some 125 000 years ago it had a position similar to that of today.

Reading for Chapter Two

The chronology and nature of the Pleistocene being so complex it is generally necessary to go to the regional literature. However, an attempt to give a general framework for Pleistocene dating and correlation is P. Evans's (1971), 'Towards a Pleistocene time-scale', *The Phanerozoic time-scale—A supplement*, Part 2, 123–356. A discussion which parallels this one is F. W. Shotton's (1966), 'Problems and contributions of methods of absolute dating within the Pleistocene period', *Quarterly Journal of the Geological Society*, 122, 357–83. A useful attempt at synthesizing the divergent views on the nature of Pleistocene chronology is provided by H. B. S. Cooke (1973), 'Pleistocene chronology: long or short?', *Quaternary Research* 3, 206–20.

The British Pleistocene is treated by R. G. West in various useful papers and books including *Pleistocene geology and biology* (1972) and 'Problems of the British Quaternary', in *Proceedings Geologists' Association of London* (1963), 74, 147–86, and he has also contributed a section to K. Rankama's (1965 and 1967) edited essays, *The Quaternary* (Interscience, 2 vols.), which bring together much material on the Pleistocene in the major European countries. More local studies on the British Quaternary include M. E. Tomlinson's (1963), 'The Pleistocene chronology of the Midlands', in *Proceedings Geologists' Association* 74, 187–202, and Penny, L. F. (1964), 'A review of the last glaciation in Great Britain', in *Proceedings Yorkshire Geological Society* 34, 387–411. A broad environmental study of the nature of the British landscape during the fluctuations of the Quaternary appears in F. W. Shotton (1962), 'The physical background of Britain in the Pleistocene', *Advancement of Science* 19, 193–206. An authoritative attempt at inter-regional correlation is by Mitchell, G. F., Penny, L. F., Shotton, F. W., and West, R. G. (1973), 'A correlation of Quaternary deposits in the British Isles', *Geological Society of London Special Report 4*, 99 pp.

Studies of the European Quaternary, additional to the useful reviews in Rankama which have already been referred to, include K. Kaiser (1969), 'The Climate of Europe during the Quaternary Ice Age', in *Quaternary geology and climate* (ed.) H. E. Wright, pp. 10–37, T. Segota (1966), 'Quaternary temperature changes in Central Europe', *Erdkunde* 20, 110–18, Wright, H. E. (1961), 'Late Pleistocene climate of Europe: a review', *Bulletin Geological Society of America* 72, 933–84.

The North American Pleistocene is discussed in a large volume of essays edited by H. E. Wright and D. G. Frey (1965), *The Quaternary of the U.S.A.* The essays are both regional and systematic in scope. The Quaternary of the Great Plains region alone is dealt with in detail in W. Dort and J. K. Jones (eds.) (1970), *Pleistocene and Recent environments of the central Great Plains*, and other essays, many of which deal with the U.S.A., appear in E. J. Cushing and H. E. Wright (eds.) (1967), *Quaternary paleo-ecology* (Yale U.P.).

Many of the publications referred to above deal with the vegetational changes of the Pleistocene, but the following papers give additional information of general value: Frenzel, B. (1968), 'The Pleistocene vegetation of northern Eurasia', *Science* 161, 637–49; Walker, D. and West, R. G. (1970) (eds.), *Studies in the vegetational history of the British Isles* (Cambridge U.P.); Turner, C. and West, R. G. (1968), 'The subdivision and zonation of interglacial periods', *Eiszeitalter und Gegenwart* 19, 93–101; Leopold, E. B. (1967), 'Late-Cenozoic patterns of plant extinction', in P. Martin and H. E. Wright (eds.), *Pleistocene extinctions* 203–46.

Broad zoological changes of the Pleistocene are described and lavishly illustrated in B. Kurten's (1973), *The Ice Age*.

A general, readable, and simple treatment of all aspects of the Pleistocene in Britain, with a particularly useful chapter on the Last Interglacial, is *The Ice Age in Britain* by B. W. Sparks and R. G. West (1972). Finally, many of the new developments on ideas of palaeoclimatology and anthropology in the Middle Pleistocene are gathered together in a massive volume, *After the Australopithecines*, edited by K. W. Butzer and G. L. Isaac (1975).

3 Pleistocene Events in the Tropics and Subtropics

> During a Pleistocene glacial advance more than half of
> the world, where man could live, was in what is now
> Africa, and Europe in fact, for most of the Pleistocene,
> must have been a small, cold, peripheral area.
>
> J. D. CLARK (1975, p. 180)

Arid phases in the Pleistocene

The events which led to the expansions and contractions of the great ice sheets during the Pleistocene also led to major environmental changes in lower latitudes. The positions of the major climatic belts were altered, and with them the major vegetation zones. One of the most striking and important results of such change was that the limits of the world's great tropical and subtropical sand deserts shifted.

One of the most satisfactory ways to assess the former extent of desert areas during the Pleistocene interpluvial or dry phases is by studying the former extent of major tropical and subtropical dune fields as evidenced by fossil forms, often visible on air or satellite photographs.

Indications that some dunes are indeed fossil rather than active are provided by features like deep-weathering and intense iron-oxide staining, clay and humus development, silica or carbonate accumulation, stabilization by vegetation, gullying by fluvial action, and degradation to angles considerably below that of the angle of repose of sand—normally 32–33° on lee slopes. Sometimes archaeological evidence can be used to show that sand deposition is no longer progressing at any appreciable rate, whilst elsewhere dunes have been found to be flooded by lakes, to have had lacustrine clays deposited in interdune depressions, and to have had lake shorelines etched on their flanks.

Sand movement will not generally take place through aeolian activity over wide areas so long as there is a good vegetation cover, though small parabolic (hairpin) dunes are probably more tolerant in this respect than the more massive siefs (linear) and barchans (crescentic). Studies where dunes are currently moving and developing suggest that vegetation only becomes effective in restricting dune movement where annual precipitation totals exceed about 100 to 300 mm. These figures apply for warm non-coastal areas. Some opinions of workers from some major desert areas on the rainfall limits to major active dune formation are summarized in Table 3.1.

At the present time overgrazing and other human activities on the desert margins may induce dune reactivation at moderately high precipitation levels, and this is, for example, a particular problem in the densely populated Thar Desert of Rajasthan, India.

Table 3.1
The rainfall limits to dune field formation

Source	Location	Today's precipitation limit for formation of moving dunes (mm)	Today's precipitation limit of fossil dunes (mm)	Dune shift in km
Hack (1941)	Arizona	238–54	305–80	—
Grove (1958)	W. Africa	150	750–1000	—
Flint and Bond (1968)	Rhodesia	300	c.500	—
Mabbutt (1971)	Australia	100	—	900
Goudie *et al.* (1973)	India	200–75	850	350
Grove and Warren (1968)	Sudan (Qoz)	—	—	200–450
Price (1958)	Texas	—	—	350
Goudie *et al.* (1973)	Southern Kalahari	175	650	—
Glassford and Killigrew (1976)	Western Australia	⟨200	⟩1000	800

When one compares the extent of old dune fields, using the types of evidence outlined above, with the extent of currently active dune fields, one appreciates the marked changes in vegetation and rainfall conditions that have taken place in many tropical areas. This is made all the more striking when one remembers that decreased Pleistocene glacial temperatures would have led to reduced evapo-transpiration rates, and thus to increased vegetation cover. This would if anything have tended to promote some dune immobilization. Dune movement might, however, have been accentuated by apparently higher trade-wind velocities during glacials (Parkin and Shackleton, 1973).

The fossil dunes of northern India
Some of the early British geologists of the Indian Geological Survey appreciated that many of the dunes in northern India were fossil forms.

As early as the 1880s W. T. Blanford remarked that in Rajasthan 'many of the sand-hills are evidently of great antiquity; despite the small rainfall of the desert region, they show signs of considerable denudation in parts, and are cut into deep ravines by the action of water'. Subsequently fossil dunes have been identified both in the Las Belas Valley area of Pakistan, in Gujarat, and in Rajasthan (Verstappen, 1970). In Gujarat the dunes, showing calcification, deep gullying and marked weathering horizons, are normally overlain by large numbers of small microlithic tools, suggesting that there has been relatively little sand movement since mesolithic man lived in the area. The fossil dunes include parabolics, transverse, longitudinal, and wind-drift types, and are now known to extend as far as Ahmedabad and Baroda in the south, and to Delhi in the east (Goudie *et al.*, 1973) (Fig. 3.1).

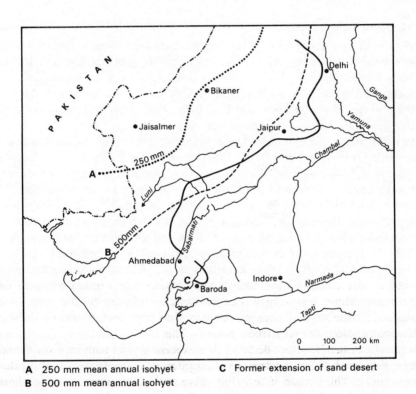

A   250 mm mean annual isohyet   C   Former extension of sand desert
B   500 mm mean annual isohyet

FIG. 3.1. The former extension of the great Indian Sand Desert in the Late Pleistocene.

They occupy zones where the rainfall is now as high as 750–900 mm. In the Sambhar salt-lake area near Jaipur in eastern Rajasthan the dunes are overlain by lake deposits of a freshwater type. The base of this lake bed has been dated at about 10 000 years B.P., suggesting that dune movement may have ceased by that time (Singh, 1971).

In numerous parts of this desert zone there appear to have been at least two major phases of aridity. The first phase seems to have been ended by red-soil development during Middle Stone Age times, and the second phase to have been ended by a period of calcification in mesolithic times. It seems possible that one reason for the relative scarcity of Upper Palaeolithic remains in this part of India may be the arid conditions pertaining between Middle Palaeolithic and Microlithic times, and this supposition is supported by the finding of an Upper Palaeolithic site within a dune near Baroda. The great profusion of microlithic sites, on the other hand, seems to be associated with the Early Holocene amelioration in rainfall conditions. A map of the approximate distribution of fossil dunes in northern India is shown in Fig. 3.1.

The fossil dunes of Africa

A basically similar picture to the Indian one comes from southern Africa, where the Kalahari sandveld of Botswana and adjacent territories is dominantly a fossil desert, now covered by a dense mixture of *acacia* and *mopane* forest, and grassland with shrubs. A detailed airphoto study has established the presence of wide expanses of Late Pleistocene dune fields in areas where the rainfall reaches 500 mm per year (Grove, 1969). They cover much of Botswana and extend into Rhodesia, where they occur in a much degraded form (Flint and Bond, 1968). Some of the ridges in north-western Botswana, to the west of Okavango, have similar morphology, though very greatly reduced in height by slope-wash, to the remarkable live parallel 100 m dunes of the hyper-arid Namib Coast of the present day. These rolling longitudinal dunes extend northwards into the Caprivi strip and, according to satellite photos, as far north as 16 °S. in Angola and Zambia. Even older dune fields extend well into the Congo rainforest zone, and reach north of the Equator (Fig. 3.2). These have become silicified into silcretes and have been greatly reworked by fluvial and soil processes. Further evidence for formerly greater humidity in the Kalahari is afforded by the long fossil valleys, locally called *mokgacha*, such as the Okwa and the Groot Laagte. There are also some palaeolake basins (see p. 79).

North of the Equator the fossil dune fields extend south into the savannah and forest zone of West Africa, and have covered lateritic and other soils and invaded palaeolake basins (Fig. 3.3). The so-called ancient erg of Hausaland extends into a zone where the present rainfall is as high as 750 mm (Grove, 1958). Many of the dunes in northern Nigeria are now cultivated, and in the vicinity of Lake Chad dunes have been flooded by rising lake waters. At one stage of the history of this area they blocked or altered the course of the Niger River. Indeed, in the middle Niger area there appear to be several ages of fossil dune, including old deeply-weathered linear dunes (*alabs*), and younger grey-brown and yellow dunes of lesser height.

The river Niger, as we now see it, was born of two parents. In the Late Pleistocene the lower, south-east flowing section was fed from the southern slopes of the Ahaggar Mountains, by affluents which are now practically extinct. Lower down it was augmented, as now, by the Sokoto and Benue.

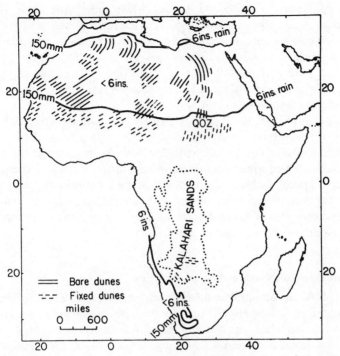

FIG. 3.2. The past and present extent of blown sand in Africa. Moving sand is confined mainly to the very gentle slopes of basin-shaped areas with mean annual rainfall totals of less than about 150 mm. Old dunes extend into more humid areas and indicate that the deserts have at times been more extensive (after Grove, 1967, Fig. 7).

The Upper Niger, flowing north-eastwards from the mountains of the Guinea–Sierra Leone border, flowed during the Late Pliocene and Early Pleistocene westwards into the Gulf of Senegal. A subsequent dry period produced a barrier of sand dunes (the Erg Ouagadou), which then blocked the previous westward flow of the upper river when the last major wet phase arrived some 10 000 to 15 000 years ago (Beadle, 1974, p. 125). It was therefore diverted into a closed basin, Lake Araouane. The flooded basin later began to drain away (either by a break-through or a capture) and, with a near right-angled turn joined the lower Niger, possibly only 5000–6000 years ago.

Further east, in the Sudan, west of the White Nile, a series of fixed dunes known locally as *Qoz* covers most of the landscape up to the slopes of the Jebel Marra. The fixed dunes extend as far south as 10 °N., and merge northwards, locally, with mobile dunes at about 16 °N. They succeed in crossing the Nile, which thus probably dried up at the time of their formation. Again, as in West Africa and India, there appear to have been at least two phases of dune activity. The first produced so-called *Low Qoz*, comprising many *alab* dunes. The second phase, which does not seem to have extended so far south, consists of more transverse dunes, and forms the *High Qoz*. These two phases were interrupted by a relatively wet phase when extensive weathering and degradation took place. The first phase suggests a shift in the wind and rainfall belts of about 450 km southwards, and the second phase of dune building in Holocene times represents a shift of about 200 km (Grove and Warren, 1968).

Periods of aridity of the type indicated by the fossil dunes of Africa did, in marginal areas, have a marked effect on man, as witnessed by the clear hiatuses that exist in the archaeological record. As Wendorf *et al.* (1976, p. 113) remark, 'There are no traces anywhere in the Nubian Desert of any occupation, spring, or lacustrine deposits that are between the Aterian sites and the Terminal Palaeolithic in age. For this period of more than 30 000 years' duration the Western Desert of Egypt was apparently devoid of surface water and of any sign of life.'

The fossil dunes of North and South America

In the U.S.A. a comparable development of fossil dune fields has been recognized. Parts of the High Plains, for example, which are now dominated by a calcrete (caliche) caprock and numerous depressions, were formerly covered by large dunefields displaying the characteristic anti-clockwise wheelround features of the dune systems of Australia and southern Africa (Price, 1958).

This American system includes the Rio Grande Delta erg which extends about 150 km from Punta Penascal at the mouth of Baffin Bay to Oilton (Torrecillas), and about 300 km from Oilton to the southern end of the Delta. Another ancient American erg is called the Llano Estacado field, and this is outlined at least in part in the present landscape by the topographic grain of etched swales, remnant ridges, and deflated swale ponds and lakes, the latter being orientated along the swales. It is often termed as 'scabland'. It has sometimes been suggested that the lineation of the dunes and lakes indicate a former wind-pattern diverging as much as 90° from the present pattern, in addition to more arid conditions. These features probably indicate an expansion of the desert to the north and east by the order of 320 km. In Nebraska and South Dakota, the Sandhills, covering an area of 52 000 km² were also more active in the Late Pleistocene, and three generations of dune formation have been recognized, the most extensive being of pre-Woodfordian (see p. 53) age (Smith, 1965). The famous Carolina Bays may have developed as deflation hollows in interdune swales.

In South America Tricart (1974) has used miscellaneous remote-sensing techniques which have enabled him to identify two ancient ergs. One was in the Llanos of the Orinoco river, where fossil dunes, partly fossilized by Holocene alluvium, extend southwards as far as latitude N.6° 30′ and 5° 20′. The other erg was in the valley of the lower-middle Sao Francisco River in

Bahia State, Brazil. At the time of its formation the river had interior drainage and did not flow throughout its length. It is also likely that aeolian activity was also much more extensive in the Pampas and other parts of Argentina.

The fossil dunes of Australia

In Australia, fossil dunes, associated with an anti-clockwise continental wheelround, have also been commented upon in the literature. They are particularly well displayed in the Pindan area of Western Australia, and in the country to the south of the Barkly Tableland. They are completely vegetated in this latter area, have subdued and rounded forms, and appear broadly comparable to those of northern Nigeria. They probably represent a decrease of rainfall in the Barkly Tableland area of between 150 and 500 mm, representing an equatorial shift of the isohyets by about 8 degrees of latitude—around 900 km—(Mabbutt, 1971). Pollen-analysis of a core at Lynch's Crater, north-east Queensland, indicates a change from a less mesophytic to a more mesophytic, regional vegetation, from the Late Pleistocene to the Early Holocene, and is convincing additional evidence for Late Pleistocene aridity in tropical Australia (Kershaw, 1974). An excellent account of the Australia fossil dunes and their interpretation is given by Bowler (1976), while Wyrwoll and Milton (1976) and Glassford and Killigrew (1976) give convincing accounts of their development in Western Australia.

Pluvial phases in the Pleistocene

No less dramatic than the evidence presented for aridity by the fossil dune systems is the evidence presented for increased hydrological activity, either resulting from a temperature decrease or an absolute precipitation increase, in the Pleistocene and Early Holocene. Such phases have been called lacustral or pluvial phases.

However, in some respects the evidence is more equivocal than that provided by the dunes, for the relationship between rainfall and lake levels is itself complicated by both temperature and non-climatic factors. With respect to the latter, it needs to be remembered that many lakes occur in areas of tectonic instability or volcanic activity, including the many lakes that occupy the floor of the East African Rift between the Danakil Depression in Ethiopia and Lake Malawi. Other lakes, including the Etosha Pan of South West Africa and the Makarikari and Ngami lakes of Botswana, may have been affected by the fact that the river systems in this semi-arid area are to a degree interconnected. Elsewhere the outlets of lakes may have been affected by erosion or vegetation growth at outlets, creating alternating lowering or ponding back of lake waters.

Nevertheless, the widespread nature and similar chronologies of many large basins in many parts of the world suggest that the climatic factor has perhaps been dominant in controlling lake level fluctuations.

Both in terms of area and depth many of these lakes were extremely prominent features of the Pleistocene environment (Table 3.2), and they were in many areas favoured sites for occupation by early man.

Table 3.2
Pluvial lake dimensions

Lake	Location	Area (km²)	Depth (m) (height above present-dry bed or lake level)
Bonneville	U.S.A.	51 700	335
Searles	U.S.A.	—	213
Panamint	U.S.A.	—	274
Russell (Mono)	U.S.A.	—	233
Lahontan	U.S.A.	22 442	213
Dead Sea	Israel	—	433
Tuz Golu	Turkey	—	75
Lake Van	Turkey	—	60
Izmik	Turkey	—	55
Burdur	Turkey	—	95
Kharga	Egypt	—	100
Dieri	Australia	104 000	46
Makarikari	Botswana	34 000	45
Nawait (Victoria and Bonney)	Australia	21 000	—
Aral–Caspian	U.S.S.R.	1 100 000	76

(From data in Butzer (1972), Flint (1971), and Grove (1969)).

The North American pluvial lakes
The greatest concentration of pluvial lakes in the western hemisphere, and possibly also in the world, occurs in the Great Basin in the northern part of the Basin and Range Province of the United States (Fig. 3.3). Between 110 and 120 depressions, formed dominantly by late Pliocene and Pleistocene high-angle faulting, were occupied wholly or in part by Pleistocene pluvial lakes. Some of these lakes, especially Bonneville (made classic by the work of G. K. Gilbert), Lahontan (Broecker and Kaufmann, 1965), and the Lake Russell to Lake Manly system, were colossal (see Table 3.2). Bonneville at its maximum extent was almost the size of Lake Michigan, but now it is occupied by only 2600–6500 km² of water. Some of the former linkages of these lakes can still be implied from their having fish species in common (Miller, 1946).

South and east of the Great Basin, in the Basin and Range Province, there were fewer depressions because of a lesser degree of Pleistocene deformation, while further south the amplitude of climatic change in the pluvials seems to have been reduced somewhat because of both increased distance from westerly storm tracks, and the greater aridity resulting from the higher, mean annual temperatures of the southerly latitudes. Nevertheless, there were pluvial lakes in these southerly areas, especially in the Mexican

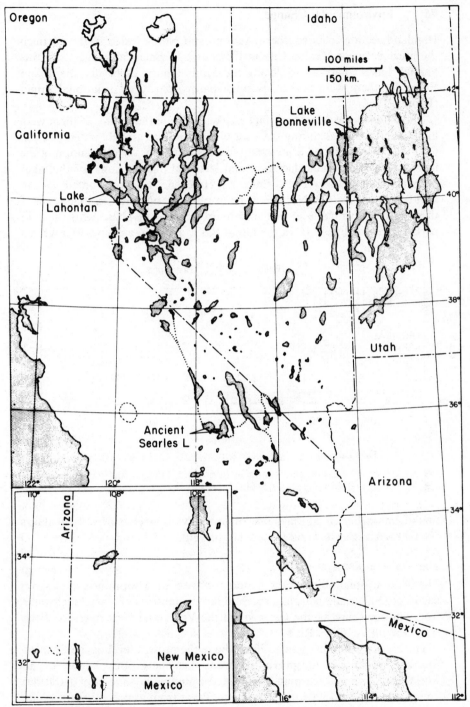

Fig. 3.3. Sketch map showing Pleistocene pluvial lakes in western United States. Dotted lines represent overflow stream channels (after Flint, 1971, Fig. 17.3).

Highland section from southern Arizona and New Mexico in the north to the great basin of Mexico City in the south. In Baja California there was Pluvial Lake Chapala, while on the High Plains, especially the Llano Estacado, there were numerous small basins, partly of deflational origin (Reeves, 1966).

The dates of the American pluvial lakes indicate that many of them were high at certain times during the Last Glacial, and as already noted (see pp. 6–7) the western United States was the classic area for the recognition of the link between mountain glaciation and lacustral phases. Some of the lakes maintained a high level into the early Holocene, a feature noted also in Africa (see p. 82). The North American pluvials, however, in that some of them coincide with the glacial maximum, are out of phase with those in parts of Africa (Fig. 3.4). In the Great Basin of Utah, a deep core revealed

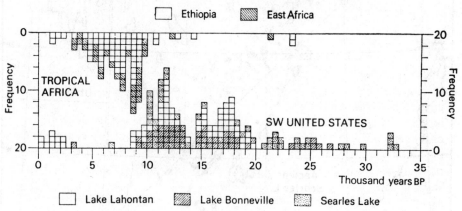

FIG. 3.4. Carbon 14 dates of lake deposits in East Africa, Ethiopia, and south-west United States (after Grove *et al.*, 1975, Fig. 7).

that there had been at least five major pluvials over a period of about 800 000 years (Eardley and Gvosdetsky, 1960).

The Aral–Caspian system

The Aral–Caspian–Black Sea system, formed in several broad shallow basins of Quaternary warping, received large quantities of glacial meltwater from various sources—the Caspian via the Volga and Ural rivers, and the Aral Sea via the Oxus River.

The highest shoreline was at 76 m above the present level of the Caspian Sea, and the area of the pluvial lake was then the greatest known in the world. The Aral and Caspian Seas united to inundate an area of 1 100 000 km², and extended 1300 km up the Volga River from its present mouth. The Caspian was also united with an expanded Black Sea through the Mantych Depression (Fig. 3.5).

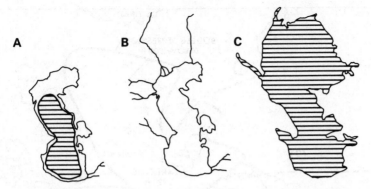

FIG. 3.5 Alternations in the extent of the Caspian Sea during the changes between warm and cold periods.
A. Area during an interglacial such as the Eemian/Mikulino.
B. Area at the present day.
C. Area during the Last Glacial (Valdai) (after Frenzel, 1973, Fig. 94).

The pluvial lakes of the Middle East

The present Dead Sea Rift Valley is currently a relatively dry terrain and contains three main lake basins, Lake Hula (now drained), Lake Tiberias (or the Sea of Galilee), and the Dead Sea itself. In Quaternary times a much greater area than today was covered by lakes, and in particular one great lake, the so-called *Lisan Lake*, extended continuously from the south shore of the present Lake Tiberias to a point some 35 km south of the south shore of today's Dead Sea. The north to south dimension of this lake was probably around 220 km, its maximum width about 17 km, and its highest shoreline was at -180 m compared to the -400 m of the Dead Sea as it exists now. In all, the water volume must have been 325 km^3 compared to the present Dead Sea with its volume of 136 km^3 (Farrand, 1971).

Tectonic disturbance of a grand scale may have been partly responsible for its decline in volume, but a lake with a shoreline at -370 m, and the altitude of its floor much the same as that of the present Dead Sea, could have held only about half the water of the former Lake Lisan, so that some climatic influence must also be invoked to explain the contraction to present dimensions. Begin *et al.* (in Rognon, 1976) believe that the Dead Sea was a substantial lake from 18 000 to 12 000, a time of great dryness in East Africa.

In the Arabian desert other old lakes have been recognized, and radiocarbon dates suggest two lacustral periods: 36 000–17 000 and 9000–6000 B.P. (McLure, 1976).

The pluvial lakes of Africa

One of the largest and most spectacular of the pluvial lakes is that of Lake Chad (Fig. 3.6), but unlike the Caspian–Aral system it did not receive glacial meltwater. At more than one stage during the Pleistocene, Chad was

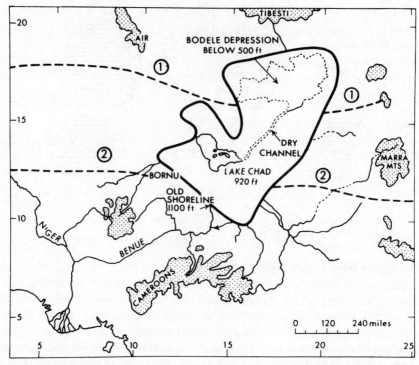

Fig. 3.6. Shifting lake shores and desert limits in the Chad basin, west central Africa. The present extent of Lake Chad, 920 feet above sea-level, is compared with the old shorelines of the ancient MegaChad at 1100 feet above sea-level (thick line), which overflowed into the Benue. Dashed line (2) is the southern limit of old, vegetation-covered dunes. It lies far south of the southern limit of moving dunes of the present day, marked by dashed line (1). These extreme changes of desert limits and lake levels probably occurred in the period 20 000–5000 years ago (after Grove, 1971, p. 7).

considerably larger than at the present time. Chad at present stands at a height of 282 m above sea-level, but at some early stage the Chari river formed a 40 000 km² delta in association with a lake at 380–400 m. The lake then shrank during an arid phase of dune formation, but later again rose to 320–30 m, and formed a marked ridge or ridge complex, traceable over a distance of more than 1200 km. Between Maiduguri and Bama in north-east Nigeria this strandline is easily identifiable as a sand mound 12 m high. This pluvial stage has been dated at about 10 000 years B.P. or a little later, though it seems likely that the lake stood at the 320 m level as late as 5000 years ago (Grove and Warren, 1968).

Diatomaceous lake deposits have been described from several widely scattered oasis depressions in the Niger Republic, and many of them possess dates of less than 10 000 years B.P.

To the east, the rift valleys of East Africa are occupied by numerous lakes around which occur Late Pleistocene and Early Holocene high strandlines.

In Ethiopia, one of the biggest of the pluvial lakes, first recognized by the Scandinavian explorer and scientist, Nilssen, is pluvial Lake Galla. This occurs to the south of Addis Ababa and is now occupied by four shrunken remnants, Ziway, Langano, Abiyata, and Shala. However, when the lakes were larger, standing as much as 112 m above the present surface of Shala, the basin was occupied by one large sheet of water (Fig. 3.7A, B) (Grove *et al.*, 1975). In Afar, Lake Abhé attained a surface area of 6000 km² and a depth of more than 150 m (Gasse, in Rognon, 1976).

Further south, the other lakes also display old strandlines and lake sediments. Lake Awasa shows a series of terraces cut in volcanic debris at 10, 22, 33, and 40 m above the present lake surface, and lakes Margherita and Chamo have a 20–30 m terrace. Oyster (*Etheria*) shells have been found at 52 m above the present swamp level. Lake Stefanie, discovered as late as 1888, is situated just to the north of the Kenya border, and increased discharge down the Sagan River appears to have led the lake, now either completely dry or seasonally flooded, to have reached a level of at least 20 m, creating fossil spits, and depositing incrustations of algal lime-stones on the old cliffs and islands (Grove, Street and Goudie, 1975, p. 183).

Further to the south similar evidence of high lake stands has been found in Kenya and Tanzania, and the chronology seems to have been similar to that in Ethiopia (see Fig. 3.4). The Nakuru–Naivasha basin had a greatly expanded lake (Fig. 3.7C).

In southern Africa relatively little research has been done on the pluvial lakes, and few dates are as yet available. However, airphoto reconnaissance does again indicate that some of the basins were very considerably enlarged in the not too distant past. In the northern part of the Kalahari sandveld in

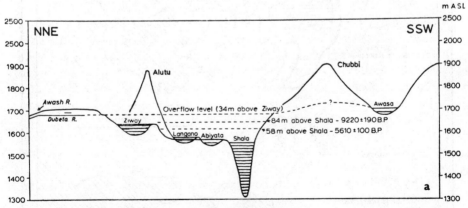

FIG. 3.7. Late Pleistocene and Holocene lakes in East Africa.
   3.7. A. NE–SW. section through the Galla lakes basin, Ethiopia, illustrating the high early Holocene levels which united Ziway, Langano, Abiyate, and Shala (after Grove and Goudie, 1971, Fig. 2).

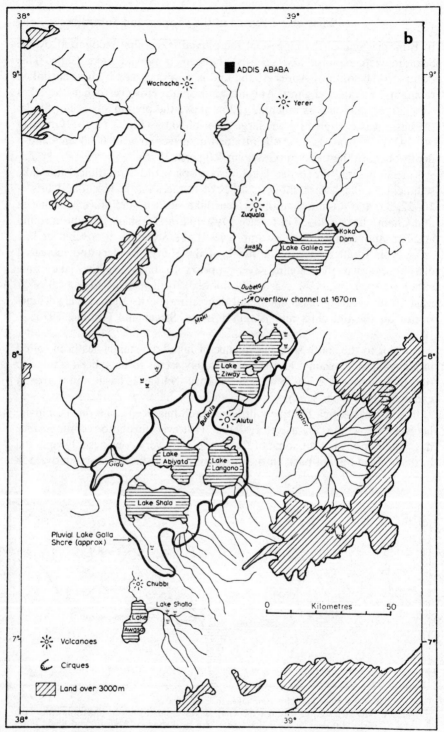

Fig. 3.7. B. The Galla lakes basin showing the pluvial lake shorelines (after Grove, Street, and Goudie, 1975, Fig. 2).

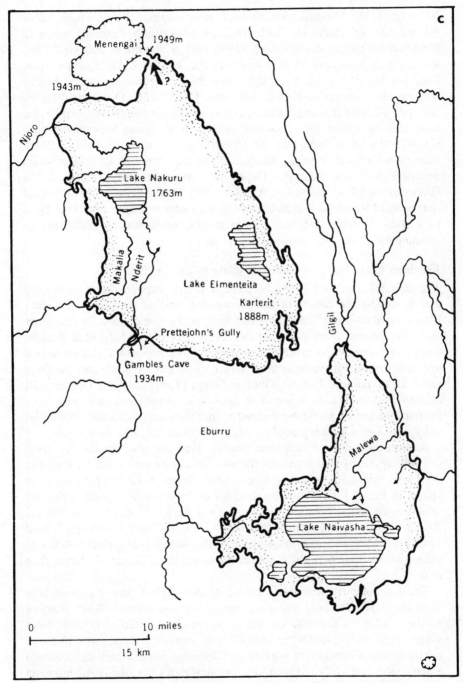

FIG. 3.7. C. The expanded Naivasha–Nakuru basin of the Early Holocene (after Butzer *et al.*, 1972, Fig. 2).

Botswana is the Makarikari depression, now occupied by the small, saline Ntetwe Pan, and Lake Dow. Pluvial Lake Makarikari, however, seems to have had an area of no less than 34 000 km², a volume of 500–1000 km³, and a depth of around 45 m. Further to the west, in Lake Ngami's basin, there are also fossil shorelines indicating that this currently small, shallow lake may have exceeded 1040 km² in area. The Mababe Depression, on the east side of the Okavango Swamps, was also occupied by a considerable lake, and its western rim is clearly defined by the great Magwikwe sand-ridge feature. It is linked to the Gubatsa Hill, and is probably a fossil sandspit or related form, as much as 20 m high, created by a lake which occupied both the Mababe Depression, and the lower parts of the Okavango and Chobe swamps (Grove, 1969). David Livingstone suspected that parts of Rhodesia and Botswana, as they now are, were once inundated by a lake—he remarked on the large number of freshwater shells that he encountered during his missionary journeys.

The dates of the last great lacustral phase in East Africa

The great lakes of East Africa, as already shown, expanded and contracted greatly during the course of the Pleistocene, and as seen on p. 148 they have even shown marked changes of level in the last decade. In that early man, like present man, occupied the lake basins of the Rift, their fluctuations have some importance. Many of the most important archaeological and anthropological finds in East Africa, as at Lake Rudolf and the Omo valley (Ethiopia and Kenya), Olduvai Gorge (Tanzania), and Olorgesaillie (Kenya), occur in association with lake beds. Another reason why these fluctuations are of particular interest is that they appear to have been partially out of phase with glacial events in northern latitudes (see p. 84).

Radiocarbon dates which have become available over the last ten years show fairly clearly that nearly all the lakes reached a maximum level around 9000 B.P., that is in early post-glacial time. Some of the available data are shown in Fig. 3.4. Comparative data for other parts of tropical Africa are included. In general there appear to be few high lake deposits in tropical Africa with dates between 18 000 and 12 000–13 000 B.P. Between about 12 000 or 12 500 and 7000 years ago, possibly with a peak around 9000 B.P., lakes were for long periods, if not for the whole time, higher and larger than now.

The dates of early pluvial or lacustral phases are less clear, partly because there are relatively fewer dates, and partly because beyond 40 000 years or so the validity of radiocarbon dating is greatly reduced. Available dates suggest that around 40 000 to about 20 000 years ago there was a phase of relatively moist conditions both in the northern Sahara and the southern Sahara (Rognon, 1976). There are very few dates for pluvial phenomena until the very Late Glacial or Early Holocene. This does imply that the glacial maxima of the Last Glaciation, dated about 23 000 to 11 000 B.P., were times of relative drought in much of Africa north of the Equator. Equally, the wet phase from 40 000 to about 20 000 may correspond very roughly with the interstadial period encountered in many parts of the northern hemisphere (see p. 53). In the Rudolf Basin of Kenya and Ethiopia, however, the last dry phase seems to have been of greater duration, 35 000–10 000 B.P., and such a dry phase is also recognized in the sequence from Lake Nakuru in Kenya.

The Australian evidence (Bowler, 1976) seems to confirm this general picture. From about 40 000 to 25 000 B.P. lake levels were high, and dunes relatively stable. After 25 000 the last major arid phase began, causing lake levels to fall. Increasing salinities assisted in the early construction of clay-rich dunes. The peak of aridity in Australia occurred around 18 000 to 16 000 B.P. (approximately the same time as the glacial maximum), when gypsum and clay dunes were constructed on the eastern margins of lakes simultaneously with expansion of desert linear dunes. By about 13 000 B.P. aridity was declining, and the dunes became stabilized.

The glacial-pluvial problem

One of the greatest problems of the Quaternary is to assess the question of what desert areas were like during glacial phases. Were they subjected to pluvials in glacials or interglacials? As already noted (see p. 7), the classic model sees glacials and pluvials as being synchronous, the shorelines of pluvial lakes appearing, in the western U.S.A., to be tied in with glacial moraines.

In East Africa the work of Wayland, Nilsson, and L. S. B. Leakey established the existence of a variable number of pluvials. Leakey's work showed that there were four main pluvials, Kageran, Kamasian, Kanjeran, and Gamblian. These were, he thought, followed by two post-pluvial wet phases, the Nakuran and the Makalian. For many people these four main pluvials, on the basis of palaeontological and archaeological evidence, were seen as being broadly correlated with the four classic glaciations of the Alps proposed by the Penck and Brückner model, and the sequence became applied almost throughout the African continent.

There are various reasons for doubting and even rejecting this simple model (Fairbridge, 1970). There are three theoretical points which suggest that full glacials may in fact have been drier: a eustatic drop in sea-level would lead to greater continentality of climate and thus greater aridity; a drop in sea-level and an extension of sea-ice would lead to less evaporation from the ocean surface, leading to less rain; the cooling of the oceans by an average of about 5 °C would lead to less evaporation and less cyclones, with the same result (Wyrwoll and Milton, 1976). Moreover, reservations produced on these grounds are substantiated in certain areas by sedimentological and geomorphic evidence. First, during the Last Glacial, siltation occurred in the middle courses of many great tropical rivers like the Nile,

Senegal, Indus, Ganges, and Narbada, and the rivers appeared to have had insufficient discharge to move their load. Secondly, deep-sea cores taken from the South Atlantic off Brazil contain much more feldspar (25–60%) in the Late Pleistocene than in the Holocene (17–20%) suggesting that chemical weathering processes were less intense in the Late Pleistocene, possibly because of a reduction in rainfall. Moreover, in South America pediments and calcretes appear in anomalously wet areas at the present day, suggesting that they too may have formed under rather more arid conditions. Thirdly, recent radiocarbon dates and chemical studies of tropical lakes indicates that they were often very dry during the Late Pleistocene, and that they frequently reached their highest levels (see p. 82) not at the maximum of the glaciations, but in late glacial and post-glacial times.

Lake Victoria appears to have been actually saline until 12 500 B.P., moderately fresh until 10 500 B.P., dry again from 10 500 B.P., and then wet from 9500 to 6500 B.P. (Kendall, 1969). Likewise fossil dunes in many desert areas can be related to the Last Glacial. In India this has been done by archaeological means, in Africa by the relationship of dunes to dateable lake deposits, and in the case of Senegal, to low sea-levels, whilst in Australia it has been noted both in New South Wales and in Western Australia that dune fields can be traced beneath estuarine muds, suggesting that they were active during low stands of sea-level, supposedly of a glacio-eustatic nature. Likewise, in the Persian Gulf and the Gulf of Oman, recent submarine research has indicated the presence of seif dune remnants present on the sea-floor (Saarnthein, 1972). They pre-dated the Holocene transgression and have been used as evidence of the Late Glacial aridity.

Another form of deep-sea core evidence that has been brought to bear on this problem is the isotopic composition of planktonic Foraminifera from the Red Sea and the Gulf of Aden. Their study (Deuser et al., 1976) indicates that during periods of maximum continental and polar glaciation in the Late Pleistocene, the Red Sea was subject to strong evaporation. Between glacial maxima, the salinity of the Red Sea was equal to or below that of the open ocean. This suggests that high-latitude interglacial periods coincided with pluvial stages in the area.

An attempt at collating some available radiometric dates for dry phases of the Late Pleistocene is made in Table 3.3. These largely confirm the evidence provided by the equivalence of low sea-levels and dune formation. Indeed the dates presented are for fossil dunes, with the exception of Colombia, where the dates are for a marked vegetational change, and for Galapagos and Lake Nakuru (Kenya), where the data are obtained from lake cores and shoreline records.

The solution to this glacial–pluvial problem may basically be a locational one, for it is likely that the shifting of wind-belts and pressure systems would tend to make some places relatively wetter, whilst it would make others relatively drier. This can be seen even at the present day. In the

Table 3.3
Dates of arid phases (interpluvials) in the Late Pleistocene

Source	Location	Date (*yrs.* B.P.)
Singh (1971)	Rajasthan (India)	pre 9 250
Michel (1968)	Senegal	post *c.*30 000
Williams and Polach (1971)	SE. Australia	pre 30 000
Twidale (1972)	Lake Eyre (Australia)	post 23 000–25 000
Butzer (1972)	Chad	*c.*22 000–12 400
Butzer (1972)	NE. Angola	*c.*38 000–12 970
van der Hammen (1972)	Colombia	21 000–13 000
Colinvaux (1972)	Galapagos	pre 24 000–10 000
Isaac, Merrick, and Nelson (1972)	Lake Nakuru (Kenya)	19 000–12 000
Bowler (1976)	Australia	25 000–13 000
Rognon (1976)	Sahara	20 000–12 000

1970s, for example (see p. 146), a run of dry years in the Sahel and certain districts of north-west India corresponded in time with a run of relatively wet years in the equatorial belt of Africa and the south of India. In the context of the Pleistocene, one approach towards an understanding of the geographical arrangement of wet and dry phases was that of Street and Grove (1976). They plotted the state of African lake levels for a series of different times for which radiocarbon dates were available, and found some striking patterns (Fig. 3.8). Thus at the glacial maximum around 18 000 years ago the northern shore of the Mediterranean was dry and dominated by an *Artemisia* steppe over wide areas (see p. 50). Thus in that area the glacial maximum coincided with a high degree of aridity. However, some radiocarbon dates for high lake levels on the north side of the Sahara (the south side of the Mediterranean basin) would suggest that in that area the reverse may have been true. Another area where the glacial maximum may also have been relatively moist was in the interior of southern Africa. Moisture-bearing climatic systems were able to penetrate further inland than they can at present as a consequence of a northward shift in the Benguela Current, and the Makarikari Lake appears to have been high at about this time. However, the tropical zones of Africa (and probably in other continents as well) witnessed low lake levels and aridity around 18 000 years ago. Pluvial conditions came to that area together with other parts of Africa around 8000 to 9000 years ago, and would appear to have been remarkably widespread.

In various areas it seems that just as there were stadials and interstadials in glacial areas so there were short pluvial or nonpluvial stages in non-glacial areas. These further complicate the simple attempts to relate glacials to dryness or wetness. Indeed, there is some support for the view that in Africa at least, pluvials were of relatively short duration (often only 2000–5000 years). There is little or no support for the view of a pluvial (or interpluvial) spanning the whole of the Last Glacial. Thus the classic East

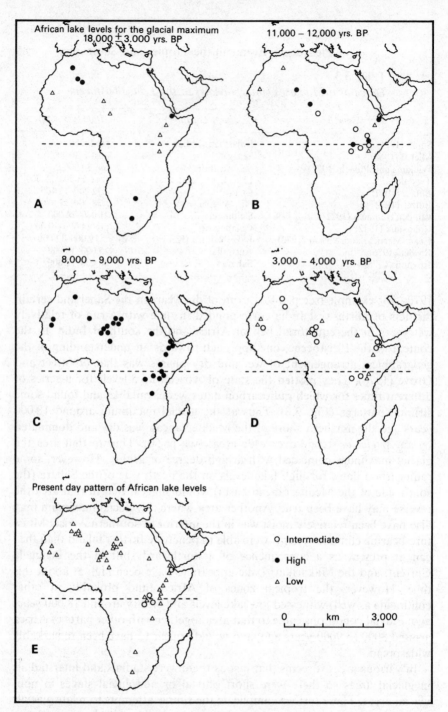

FIG. 3.8. Radiocarbon dated lake level fluctuations in Africa from *c*. 18 000 B.P. to the present (from Street and Grove, 1976).

A. At the glacial maximum, 18 0000 ± 3000 B.P.
B. At 11 000–12 000 B.P.
C. At 8000–9000 B.P.
D. At 3000–4000 B.P.
E. At the present day.

African pluvial sequence developed during the inter-war period to correspond with the European glacial sequence should be totally abandoned.

Faunal and floral changes in the tropics

The massive environmental changes described so far in this chapter have led
to changes in floral and faunal distributions in the tropics, and to curious or
anomalous patterns. The classic example of this is the distribution of the
crocodile in Africa. It was formerly ubiquitous in the rivers of that continent from Natal to the Nile. Today it is found in pools in the Tibesti Massif
in the heart of the Sahara, 1300 km from either Niger or Nile, and clearly
isolated. There is no prospect of natural migration there across the arid
Saharan wastes, given present hydrological conditions, so that pluvial conditions presumably played a role.

Another example from Africa illustrates the way in which the flora of the
East African mountains has become isolated. The distinctive tree heath
(*Erica arborea*) occurs in disjunct areas including the mountains of
Ruwenzori, the Ethiopian mountains, the Cameroon Mountains in West
Africa, and the mountains of the Canary Islands. In addition to this highly
fragmented distribution in Africa, the plant has an extensive range in
Europe from Iberia to the Black Sea. Once again it seems likely that post-
glacial changes in temperature and rainfall have led to this position. In
general, because of its height characteristics, the African continent would
have been particularly severely affected by temperature depression in the
glacials. The effect of a temperature depression of 5 °C would have been to
bring down the main montane biomes from around 1500 m to 700 or 500 m
(Moreau, 1963). Instead of occupying a large number of islands as it does
now, and as it must have done in interglacials, the montane type of biome
would have occupied a continuous block from Ethiopia to the Cape, with
an extension to the Cameroons. The strictly lowland biomes, comprised of
species today that do not enter areas above 1500 m, would, outside West
Africa, have been confined to a coastal rim and to two isolated areas inland
(the Sudan and the middle of the Congo basin).

Further massive changes in African biomes would have been occasioned
by changes of humidity as well as of temperature. In West Africa, where the
great sand ergs of the Sahara encroached as much as 500 km on the more
moist coastal regions, the southward movement of the vegetation belts cannot have failed to have had powerful effects on flora and fauna. Indeed,
since at present the West African rainforests nowhere reach inland by so
much as 500 km, if the entire system of vegetation belts had shifted south as
much as did the Saharan dunes, then the whole of the West African forests
would have been eliminated against the coastline. The present richness,
however, of the West African forests, and the existence of so many endemic
species there, makes it virtually certain that this did not happen, but the
effect of the southward advance must have been formidable. Moreau

believes that it is probable that during the arid period in which the dunes were established to the south of their present limits, the savannah on the coast, now restricted to a few small areas, stretched so far as to eliminate the forests of western Nigeria, and, joining the Dahomey Gap, produced a gap of over 1100 km between the forests of Upper Guinea and the neighbourhood of the Cameroons.

Such environmental changes, by producing new environments, and by isolating species in a restricted area, lead to the development of some species which are 'endemic', that is, which are entirely confined to an area. It also follows that the longer the period of isolation, and the more effective the barriers to dispersal, the more the local species would diverge from the original population. In extreme cases endemic genera, or even families, might evolve and be restricted to a relatively small region. Speciation is complete when the divergence of two or more portions of an original population involves the reproductive system, and thus prohibits interbreeding, so that they remain distinct and may continue to diverge, even if circumstances should bring them together yet again.

Examples of such basic evolutionary processes related to Pleistocene environmental changes in the tropics can be provided by a study of the faunas of the Amazonian rainforest and the lakes of East Africa.

In the rainforest zone of South America there are are some curious speciation patterns at the present day involving birds, trees, butterflies, and lizards (Haffer, 1969; Prance, 1973; Brown *et al.*, 1974). These seem to result from changes in the nature and extent of the rainforest in the Quaternary. Areas of what is called secondary contact have been established between distinguishable forms of birds and lizards. These are recognized as stepped clines, hybridization belts, places with character displacement, or areas with narrow sympatry of closely related fauna. In the case of Amazonia there is a striking coincidence in the location of the areas of such secondary contact between unrelated groups of birds: cracidae, tucanets, parrots, cotingids, and manakins. In most sexually reproducing, outcrossing organisms, such as birds, differentiation can only occur if populations are isolated from one another (reduced gene flow). Consequently we can assume that an area in which we find secondary contact and overlapping of differentiated forms, indicates where the two forms had previously been separated. If the modern zone coincides with a discernible physiographic feature, such as a mountain range, a large lake, a river, and so on, it is probable that the feature is, and was, a barrier to gene flow. However, in the Amazon rainforest the present area of overlap between different species does not seem to coincide with any visible physical or ecological feature. Thus one can postulate that a barrier existed in the past which is no longer operative. Recent geomorphological work indicates that during parts of the Pleistocene the great Amazonian rainforest, which at present shows a considerable uniformity over vast areas, was fragmented into small isolated patches by greatly increased aridity (Fig. 3.9). The area of savanna was greatly extended (Van der Hammen, 1974). It was these small isolated patches of rainforest, concentrated in areas of favourable hydrological conditions, which enabled the differentiation to take place in the various species of the area (Vuilleumier, 1971). The return of pluvial conditions has enabled the rainforest to spread

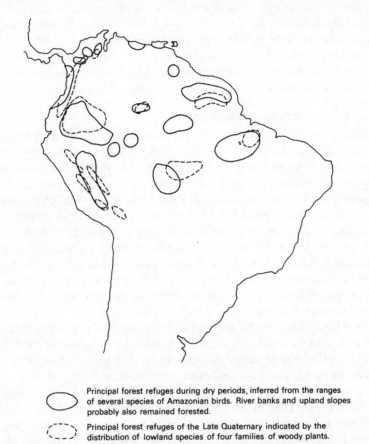

Principal forest refuges during dry periods, inferred from the ranges of several species of Amazonian birds. River banks and upland slopes probably also remained forested.

Principal forest refuges of the Late Quaternary indicated by the distribution of lowland species of four families of woody plants.

FIG. 3.9. Late Pleistocene refugia during dry periods, inferred from the ranges of species of Amazonian birds and woody plants (after Meggers, 1975, Figs. 3 and 4).

once again, and for the formerly isolated species to merge together again in zones of secondary contact.

The effectiveness of such disruption of the rainforest in leading to distinct speciation of the Amazonian forest birds would depend to a large extent on the rate at which the evolutionary process might operate. Research by Haffer (1969) suggests that under favourable circumstances the speciation process in birds may be completed in 20 000–30 000 years or less. This estimate refers mainly, thought not exclusively, to passerine birds with a high reproductive rate and evolutionary potential. If this order of magnitude is approximately correct it means that within one long interpluvial the necessary degree of speciation could take place, and that during the whole length of the Quaternary the Amazonian birds may have speciated repeatedly.

The full implications of this new work on the disruption of the rainforest by Pleistocene aridity, substantiated as it is from a remarkable variety of different sources, have probably yet still to be worked out. Already, however, anthropologists have recognized that the aridity would have posed problems for subsistence economies, and that it could have played a role in the present pattern of cultural distributions (Meggers, 1975).

African fish and changing African waters

For about forty years work has been undertaken on the environmental changes of East Africa as they might have influenced the distribution characteristics of the water fauna of the great lakes. During pluvial or lacustral phases the rivers and lake basins of East Africa would have been linked up to a greater extent than they are now. During dry interpluvial phases on the other hand the lakes would have become partly or wholly desiccated, and the linkage of water bodies would have been reduced. Such fluctuations would lead to alternations of contact and isolation for fauna, and complete lake desiccation would have led to the extermination of many species in any one basin. By a study of present and fossil fish types, also of crocodiles, the consequences of these changes can be identified, and this helps to unfathom the anomalies of zoo-geography (Beadle, 1974).

Lake Rudolf, between Kenya and Ethiopia, is now not connected with the Nile, having no outlet. On the other hand, it has a fauna very similar to that in the Nile. The explanation seems to be that Rudolf once stood at a much higher level. There are shorelines to prove this. When this was so the lake could have overflowed via a gorge, now dry, to the Sobat River, and hence to the Nile. This former connection would explain the faunal similarities. There are nevertheless, twelve endemic species of fish in Rudolf, and the Nile Perch has divided into two subspecies. The Nile Perch also occurs in Lake Stefanie, Abaya and Chamo. These too were formerly connected to Rudolf (Grove, Street and Goudie, 1975, p. 183).

A more complex situation is illustrated by Lake Kivu. Tectonic changes have led to its loss to the Nile but adoption by the Congo. Formerly, Kivu was connected with Lake Edward and the Nile via the River Ruchuru. However, in comparatively recent times this outlet for Kivu was dammed by volcanic outpourings from the Birunga Range, and so the lake increased in size and overflowed southwards towards Lake Tanganyika. It is for this reason that Lake Kivu contains some characteristic Nile fish like the barbel (*Barbus altianalis*), even though it is no longer connected with that river.

Desiccation of the lakes explains the differences between fossil and current faunas and also helps to explain the absence of some species from some of the lake basins. For example, Lake Edward in Uganda possesses no crocodiles, even though they are found in Lake Victoria, and in the Semliki River. It is difficult to explain the fact that the crocodiles have not managed to pass beyond the Semliki gorge to Lake Edward, but is probable that the falls of the Gorge, and the dense forest on either side, have acted as barriers. However, in fossil beds which border the shores of the Kazinga Channel, numerous teeth, scales, and bones of crocodiles have been found. Thus these reptiles once lived in large numbers in the lake, and so it appears likely that their present absence can be explained by the desiccation of the lake, and by the natural barriers which have prevented re-colonization. Violent volcanic eruptions may also have led to their destruction.

The recent work of Kendall (1969) and others has shown that Lake Victoria too largely dried up in the Late Pleistocene, and cores in other lake beds also indicate that many of the lakes became highly alkaline, if not actually dry, at some point. The only fish that could survive the very dry conditions of the interpluvials would be the lung-fish and the mud-fish, for they can burrow into mud and live there for long periods. It is therefore interesting that it is these two species which are most widely distributed in the Nile and the lakes at the present time, for they still live both above and below both the Murchison and the Semliki falls. Many other fish were killed off by the aridity, unlike the lung- and mud-fish, and only partial re-colonization has been possible because of the barriers presented by these two falls.

One of the most striking demonstrations of the speed with which the process of speciation occurs is provided by Lake Nabugabo, a small shallow basin of the western shore of Lake Victoria. It was disconnected from Lake Victoria about 4000 years ago by the growth of a sandbar. Three new species of *Haplochromis* have evolved during this relatively short span of time.

Elsewhere in Africa there are further zoo-geographical anomalies of interest. For instance, there are some species of fish which are common to all the major basins of the Sudan belt, Senegal, Gambia, Volta, Niger, Chad, and Nile. The fish fauna is remarkably uniform over this enormous area, which is all the more surprising when one realises that there are now more than 1600 km of desert separating the Nile from Lake Chad. This similarity may be explained by the presence of more rivers and lakes during humid phases. These would provide the necessary linkages (Beadle, 1974, p. 147). Likewise, it was a surprise to zoologists when French expeditions during the early years of this century discovered that the Sahara itself supports a considerable, though widely scattered, freshwater vertebrate fauna. In permanent but isolated water holes from Biskra to Tibesti there are the remains of a former tropical African fish fauna such as *Tilapia zillii*, *Astotilapia*

(Haplochromis), *Desfontainesii*, and *Clarias lazera*—species that are now very widespread in tropical Africa. They have become isolated as a result of a decrease in humidity (Beadle, 1974, p. 157).

Reading for Chapter Three

The reading on climatic change in the tropics is still limited in scope, and there is no one good review of work that has been undertaken. For information on the last 20 000 years the best review is A. T. Grove (1967), 'The last 20 000 years in the tropics', *British Geomorphological Research Group Special Publication* No. 5, ed. by A. Harvey. Another useful review is that by R. F. Flint (1963), 'Pleistocene climates in low-latitudes', *Geographical Review* 53, 123–9. Two other reviews of more than local interest are K. W. Butzer (1961), 'Climatic change in arid regions since the Pliocene', *Arid Zone Research* (UNESCO) 17, 31–56, and R. W. Fairbridge (1970), 'World climatology of the Quaternary', *Revue de Géographie Physique et de Géologie Dynamique* 12 (2), 97–104. The latter discusses the question of the contemporaneity or otherwise of glacials and pluvials, as does M. A. J. Williams (1975), 'Late Pleistocene tropical aridity synchronous in both hemispheres?', *Nature* 253, 617–18.

The following regional treatments give a good impression of the over-all nature of low-latitude environmental changes: K. W. Butzer (1958), 'Quaternary stratigraphy and climate in the Near East', *Bonner geographische Abhandlungen* 24 (157 pp.); R. W. Galloway (1965), 'Late Quaternary climates in Australia', *Journal of Geology* 73, 603–18; T. Monod (1964), 'The Late Tertiary and Pleistocene in the Sahara', in F. C. Howell and F. Bourlière (eds.), *Background to human evolution*, 117–229; A. T. Grove and A. Warren (1968), 'Quaternary landforms and climate on the south side of the Sahara', *Geographical Journal* 134, 194–208; A. T. Grove (1969), 'Landforms and climatic change in the Kalahari and Ngamiland', ibid. 135, 191–212; A. S. Goudie, B. Allchin, and K. T. M. Hegde (1973), 'The former extensions of the Great Indian Sand Desert', *Geographical Journal* 139, 243–57.

The bearing of recent radiocarbon dates on the history of pluvial lakes in Africa is discussed in two remarkably similar papers: A. T. Grove and A. Goudie (1971), 'Late Quaternary lake levels in the rift valley of southern Ethiopia and elsewhere in tropical Africa', *Nature*, 234, 403–5, and K. W. Butzer *et al.* (1972) 'Radiocarbon dating of East African lake levels', *Science* 175, 1069–75.

The literature on the theme of environmental change in the tropics and its effects on man and animals and vegetation is diffuse. A useful general review on the significance of the Pleistocene for current world distribution of plants and animals is E. S. Deevey (1949), 'Biogeography of the Pleistocene', *Bulletin Geological Society of America*, 60, 1315–416. Equally, B. Seddon (1971), has some useful material in his *Introduction to Biogeography*, as does J. K. Charlesworth (1957), in his massive *Quaternary Era*. On a more local scale R. E. Moreau's (1963), 'Vicissitudes of the African biomes in the Late Pleistocene', *Proceedings Zoological Society of London* 141, 395–421, E. S. Deevey's (1949), 'Living records of the Ice Age', *Scientific American* (May), A. and D. Löve's (1963), *North Atlantic biota and their history*, B. S. Vuilleumier (1971), 'Pleistocene changes in the fauna and flora of South America', *Science*, 173, 771–80, and J. Haffer (1969), 'Speciation in Amazonian forest birds', ibid. 165, 131–7 are especially valuable. Many valuable essays dealing with many aspects of the theme of this chapter in the Australian context appear in D. J. Mulvaney and J. Golson (1971) (eds.), *Aboriginal man and environment in Australia* (A.N.U. Press, Canberra). A comparable volume for Africa, but dealing of necessity with a more extended time-span is F. C. Howell and F. Bourlière (1963) (eds.), *African ecology and human evolution* (Aldine, Chicago). A fascinating discussion of how African fish populations have changed in response to Pleistocene environmental changes is by L. C. Beadle (1974), *The inland waters of Tropical Africa: an introduction to tropical limnology.*

4 Environmental Change in Post-glacial Times

> Since geological time is not salami, slicing it up has no
> particular virtue. But if it is to be sliced there is no need
> to botch the job, and chronometric dating provides the
> guidelines.
>
> C. Vita-Finzi (1973, p. 47)

A stable Holocene?

The ending of the Last Glacial period was not the end of substantial environmental change, though from time to time the existence of any major changes of climate have been doubted, and recently a hydrologist, Raikes (1967) has put forward the hypothesis that 'From at latest 7000 B.C., and possibly earlier, the worldwide climate has been essentially the same as that of today'. He believed that with the single exception of localized changes induced by large-scale, eustatic–isostatic marine transgression, all changes since about 7000 B.C. have been 'local, random, and of short duration'.

Raikes was especially sceptical about pollen, zoological, and historical evidence for Holocene climatic change, and pointed out correctly that man had influenced vegetation, that animals are often poor ecological indicators, and that the evidence for a dense population in the Indus Valley in prehistoric times (the Harappan civilization) could be explained by non-climatic factors. Nevertheless, Raikes ignores or largely ignores much of the evidence that has been put forward in many countries for Holocene climatic changes.

There are a whole series of faunal and floral remains which indicate, for instance, the relatively higher temperature conditions of the Hypisthermal interval (p. 114); there are a large number of radiocarbon dates which illustrate the fluctuations of glaciers (p. 116); there are a large number of dates of lacustrine sediments indicating Holocene lacustral phases in the tropics and subtropics (p. 111); and there are the more recent meteorological and hydrological records which indicate considerable changes and fluctuations in the last two centuries (Chapter Five). Such evidence taken as a whole shows quite clearly that the concept of a stable Holocene environment is quite untenable. This chapter is concerned with both the evidence for Holocene changes, and with the nature and influence of the changes themselves. It starts off by considering the nature and effects of the transition from a glacial to a non-glacial environment, and then it considers some of the major events of the Holocene itself.

The transition from Late Glacial times

As we have seen, the Last Glacial came to an end approximately 11 000 or 10 000 years ago, though it probably reached its maximum around 18 000 years ago. The Late Glacial phase was marked by various fluctuations, interstades of short duration, including the Erie Interstadial of the Laurentide ice sheet (around 16 000 B.P.), the Two Creeks Interstadial of the Great Lakes area (around 12 800 to 11 800), the Berezayka Interstadial (18 250 B.P.), the Ula or Somino Interstadial (16 000 B.P.) and the Raunis Interstadial (13 390 B.P.) in the U.S.S.R. (Fig. 4.1A), and the Lascaux Interstadial in France (16 000 to 17 000 B.P.). Other indications of short interstadials were provided by the Camp Century Ice Core from Greenland (13 100; 14 100 and 14 900 B.P.).

In Europe (see Table 4.1) too there were a number of distinct cool phases interrupting the retreat of the Scandinavian ice margins, and a number of

Table 4.1

The classic European Holocene sequence

Period	Zone number	Blytt–Sernander Zone name	Radiocarbon years B.P.
Post-Glacial	IX	Sub-Atlantic	post 2 450
	VIII	Sub-Boreal	2 450–4 450
	VII	Atlantic	4 450–7 450
	VI	Late Boreal	7 450–8 450
	V	Early Boreal	8 450–9 450
	IV	Pre-Boreal	9 450–10 250
Late Glacial	III	Younger Dryas	10 250–11 350
	II	Allerød	11 350–12 150
	Ic	Older Dryas	12 150–12 350
	Ib	Bølling	12 350–12 750
	Ia	Oldest Dryas	

After Embleton and King, 1967, Table 3, p. 14, and other sources.

short interstadials during which retreat was accelerated. The South Scanian, Danish Langeland, and Inner Pomeranian moraines, for example, belong to the Older Dryas (Zone I), while the Norwegian Ra, Central Swedish, and Salpaussellkä moraines belong to the Younger Dryas (Zone III). In Britain, the Younger Dryas witnessed the Corrie Glaciation in the English Lake District and in Wales, while the Older Dryas saw a re-advance of the glaciers in Scotland (the Perth–Aberdeen Re-advance). The general trend of the temperature fluctuations associated with these fluctuations is shown in Fig. 4.1 (A, B), as is the sudden increase in warmth after about 10 000 B.P.

The character, identification, and correlation of the Late Glacial interstadials is, however, a matter which is still in need of clarification. The

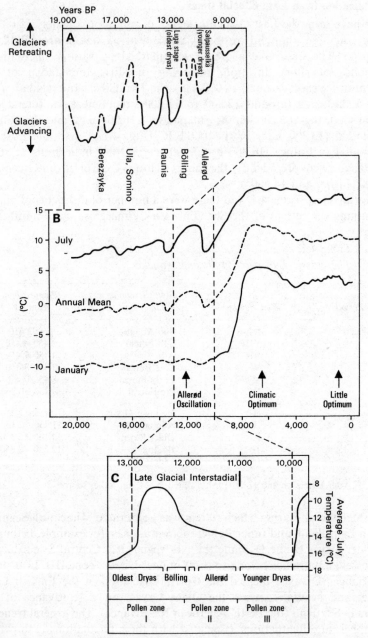

FIG. 4.1. Late glacial climatic fluctuations revealed from various sources:
  A. Glacial oscillations in the Russian Plain (after Chebotareva, 1969).
  B. Estimated trend of the summer and winter mean temperatures during the past
     20 000 years in the English Midlands (after Manley, 1964).
  C. Average July temperatures for Central England based on Coleoptera remains
     (after Coope, 1975).

classic threefold division into two cold zones (I and III) separated by a milder interstadial (II) emanates from a type section at Allerød, north of Copenhagen, where an organic lake mud was exposed between an upper and lower clay, both of which contained pollen of *Dryas octopetula*, a plant tolerant of severely cold climates. The lake muds contained a cool temperate flora including some tree birches, and the milder stage which they represented was called the Allerød Interstadial. The Interstadial itself, and the following Younger Dryas temperature reversal, are sometimes called the Allerød oscillation. A somewhat earlier minor Interstadial, the Bölling, has also been recognized in parts of Europe. On the basis of floral studies attempts have been made to reconstruct the nature of the European land-scape as it probably appeared in the Allerød Interstadial. It is worth com-paring this figure (Fig. 4.2) with that for the maximum of the Last Glaciation (Fig. 2.9). The ice sheets have been greatly reduced in extent in

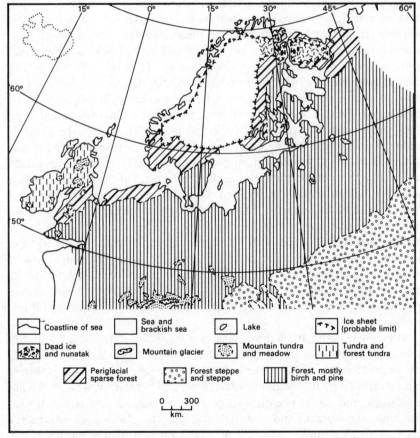

FIG. 4.2. Palaeogeographic reconstruction of northern Europe during the Allerød Interstadial (after Gerasimov, 1969, Fig. 4).

comparison with their maxima, but sea-levels are still low, Britain is connected with the Continent, Denmark is relatively unfragmented into islands,
and tundra vegetation is less widespread. However, pine woodlands are
known to have dominated the southern half of France, southern Germany,
and northern Poland, and birch to have occupied much of northern France
and northern Germany. Most of Fennoscandia was still glaciated.

Two major problems persist, however. The first is whether there is
evidence for the Allerød oscillation outside Europe. For example, while
there do appear to have been various oscillations in ice margins in North
America in Late Glacial times, it is difficult to prove their contemporaneity
with the Allerød oscillation (Mercer, 1969), though pollen-analysis in East
Africa and South America (Coetzee, 1964; Hammen, 1974) does suggest a
direct comparibility with the European sequence.

A second problem is posed by a comparison of the palynological and the
Coleopteran evidence in Britain (Coope, 1975). The Coleoptera do support
the general interpretation of a Late Glacial climatic oscillation, but they
also indicate that perhaps its thermal maximum differed both in timing and
intensity from that inferred from the floral evidence. The main phase of
warmth suggested by Coope is between 13 000 and 11 000 B.P. (Fig. 4.1C),
with the peak occurring during Zone I time. He believes that this 'Late
Glacial interstadial' cannot be correlated with the Bölling or Allerød oscillations of continental authors, since only one oscillation is indicated by the
Coleoptera and this does not correspond with either. The Interstadial began
well before, and also reached its thermal maximum before, the generally
accepted date of the Bölling oscillation (Zone Ib). Furthermore, the climate
indicated by the Coleoptera during the thermal maximum was apparently
warm enough to support a mixed deciduous forest, but the pollen spectra at
this time indicate a much more open country which has been interpreted as
having a tundra-like climate. This apparent difference between the climates
indicated by the insects and the plants may perhaps be explicable by the
greater facility and speed with which the insects would respond to rapid
climatic fluctuations in comparison to, say, birch trees. The arrival time of
birch trees in a particular area is likely to be more closely related to the
distance of the glacial refuge from which it spread when the climate
ameliorated, rather than to the time of the onset of climatic conditions
acceptable to birch.

After the Allerød oscillation, whatever its exact status, the traditional
division is made between the Late Glacial (Pleistocene) and the Post-Glacial
(Holocene, Recent, or Flandrian). The Classic terminology of the Holocene
was established by two Scandinavians, Blytt and Sernander, who, in the late
nineteenth and early twentieth centuries, introduced the terms Boreal,
Atlantic, Sub-Boreal, and Sub-Atlantic for the various environmental
fluctuations that took place. These terms are still widely used for subdivisions of the Holocene (Table 4.1). The fluctuations which Blytt and

Sernander and subsequent workers have established, though they have sometimes been contested by workers who believe the sequence of events has been less complex, with only a simple climatic optimum and then deterioration, have been remarkably durable. Nevertheless, it has to be remembered that it is essentially a scheme of vegetation change, and not a scheme of climatic change. The bulk of the evidence used by Blytt and Sernander was provided by plant remains, and especially macrofossils. Thus in terms of climatic reconstruction certain inaccuracies may creep in because of nonclimatic factors affecting vegetational associations. Such factors include the intervention of man (see p. 106), the progressive evolution of soils through time (see p. 107), and the passage from pioneer to climax species during the course of succession. As already noted with regard to the Allerød Oscillation, plants may not be able to respond with great alacrity to climatic change: migration and colonization take time. Consequently, although the terminology may still be used, there have been substantial changes in interpretation of the classic Blytt–Sernander model in recent years.

For man the time of the waning of the ice sheets brought a very great and apparently rapid alteration of the environment throughout the world. For the most part these changes did, according to Sauer (1948) and others, offer increased opportunities, though in some areas a reduction in rainfall may have reduced the possibilities of life in marginal desert areas, while increasing tree cover in northern Europe may have adversely affected the Upper Palaeolithic hunting communities.

The recession of the ice sheets uncovered millions of square kilometres of land in higher latitudes, which thus became available for human occupation and colonization. The world population of migrant waterfowl possibly increased hugely with the addition of the great breeding and feeding grounds of the northern hemisphere. Further, the transgression of the sea, resulting from the melting of the ice caps, though it may have led to a considerable inundation of the continental shelves (see p. 173) did in some ways improve the sea shores for man. A more diversified and sinuous coastline would give a wider choice of environments, while the drowning of valleys to give rias (sinuous inlets of the sea) would tend to lead to an increase in tidal ranges which would be most valuable for food-collecting peoples. The formation of quiet landlocked bodies of water might also provide a favourable setting for early trials in navigation. Many alluvial valleys grew in length and breadth and offered optimal sites for plant growth. As Sauer (1948, p. 258) has put it,

A new world took form, developing the physical geography that we know. The period was one of maximum opportunity for progressive and adventurous man. The higher latitudes were open to his colonization. In mild lands rich valleys invited his ingenuity. It was above all a rarely favourable time for man to test out the possibilities of waterside life, and especially of living along fresh water.

Environmental change and the transition from Upper Palaeolithic to Mesolithic

The transition from the Pleistocene to the Holocene also saw the transition from an Upper Palaeolithic technology to a Mesolithic and Microlithic technology. J. G. D. Clark believes (1970, p. 90) that in terms of this technological change 'Beyond a doubt the most important factors involved were the complex changes in the physical environment that marked the onset of Neothermal conditions at the close of the Ice Age and adjustments to these made by the hunter-fishers themselves.'

The main environmental change involved was the dramatic change in temperate Europe whereby forest trees were able to expand from their refuge areas and to colonize the relatively open spaces of the Late Glacial landscape. Such a change, resulting from the increasing temperatures of the Holocene, was by no means advantageous to the European hunting peoples. Indeed, the reverse may well have been the case, for the late Magdalenians and their counterparts on the North European Plain were adapted to hunting animals in a relatively open and unforested environment. This environment was one which was highly favourable to herds of reindeer, bison, and horse, and the outstanding development of certain forms—for example, the Giant Irish Deer (*Megaceros giganteus*) attained an antler span of up to 3·4 m—emphasize how suitable grazing conditions must have been. The onset of forested conditions in post-glacial times must have been little less than catastrophic. The spread of forest reduced the density of grazing animals and meant that, instead of being hunted in herds, they had to be killed individually by the hunters in the forest. This reduction in the supplies of easily killed game probably led to the intensification of methods which characterized the change from the Upper Palaeolithic method of hunting to that of the Mesolithic. The bow came into much more general use, and the microlith used to barb and tip arrows became a veritable symbol of the Mesolithic phase.

Clark (1970, p. 96) has written that

The coincidence between the passing of Late-Glacial and Late-Pluvial climate and the emergence of Mesolithic societies is more than merely temporal: it must surely have been causal, even if the precise links are not always apparent. Traditions formed under ecological conditions that had passed away had either to disappear or to undergo the modifications needed to accommodate them to new ones.

It was only in the far north of Europe, where conditions of temperature became markedly more favourable to human activities, that post-glacial warmth brought very marked advantages, so that there was in mesolithic times a dramatic expansion of people into Scotland, northern Ireland, and as far as the White Sea coasts in Norway and Finnmark.

Early Holocene environmental change and the origins of agriculture

For some considerable time it has been suggested that supposed climatic desiccation in the Near East at the end of the Last Glaciation, which has already been referred to, might have played a role in man's adoption of a food-producing economy. The archaeologist Gordon Childe (1954) expressed the opinion that 'Enforced concentration by the banks of streams and shrinking springs would entail a more intensive search for means of nourishment. Animals and man would be herded together in oases that were becoming increasingly isolated by desert tracts.' Likewise, East (1938), a historical geographer, supposed that the response of man to the desiccation of the Afrasian grasslands as storm tracks moved north could have been in four main ways: he could emigrate to more congenial areas; he could stay where he was and if he survived the hard conditions he would have to modify his life; he could think up an entirely new means of livelihood, through cultivation and animal husbandry; and he could explore the possibilities of the formerly neglected riverine lands.

Certain researches in the mountains north of Mesopotamia throw light on this relationship between early food production and environment. Wright (1968) and his co-workers have shown that the high parts of the Zagros Mountains were glaciated in Pleistocene times and that the snowline must have been 1200–1800 m lower than today. Beneath the snowline, conditions would have been cold, and the main vegetation association would have been a bleak steppe. The environment was too cold for man to live in the mountains between 28 000 B.P. and about 13 000 B.P. The environmental change from a cool steppe to a warm oak–pistachio savanna about 11 000 years ago, as determined by pollen and lake sediment studies, occurred at the same time as the first manifestations of domestications of plants and animals. Emmer and barley probably arrived in the area at this time, following the climatic amelioration, and man was able to live elsewhere than in caves. Wright (1968) has written:

Although I have always felt that cultural evolution—gradual refinement of tools and techniques for controlling the environment—is a stronger force than climatic determinism in the development of early cultures, the chronological coincidence of important environmental and cultural change in this area during initial phases of domestication is now well enough documented that it cannot be ignored. A much greater problem, of course, will be to prove that the environmental change was the cause of the cultural revolution.

Solecki (1963), on the other hand has been more bold: 'A kind of trigger was needed to make him (man) depart from being a perpetual "lotus-eater" forever dependent upon hunting and gathering for his existence. In the area under discussion the rise in temperature could have served as just such an indirect stimulus.' This theme has been investigated in greater depth by Butzer (1972).

In other parts of Asia there are also records of an end to the hiatus which separated the Mousterian and the Upper Palaeolithic from the Mesolithic. In Rajasthan (India), for example, the sudden appearance of large numbers of microliths seems to mark the end of a marked phase of Pleistocene aridity, and judging from the evidence of freshwater lakes the humid phase started around 9000 to 10 000 years B.P. (see p. 113). Also, Solecki has related how after 11 000–12 000 B.P. 'there was a rash of Mesolithic settlements'. As in the Near East, 'they blossomed over what is now Soviet Asia like desert flowers after a rain, taking advantage of an apparent cultural vacuum . . .'

The great extinction problem of Late Glacial and Early Holocene times

Another major event associated with the transition from Late Glacial to post-glacial times was the demise of many of the world's mammals. As Alfred Wallace put it in 1876 'We live in a zoologically impoverished world, from which all the hugest, and fiercest and strangest forms have recently disappeared.' The great geologists and zoologists, men like Darwin, Lyell, Owen, and Cuvier, were fascinated by this enduring problem of the great Pleistocene extinctions or 'overkill' of the earth's mammalian population (Martin, 1966). For recent workers, however, the problem goes beyond explaining the massive reduction in species, particularly of big game mammalian and avian herbivores *per se*: the concern is why the biggest wave of extinctions occurred only once, and at a time within the last 15 000 years, except, that is, in Africa and South-East Asia, where the extinction probably occurred before 40 000 to 50 000 years B.P.

Table 4.2 gives the dates for the start of the major extinctions as proposed by Martin (1967).

The real cause of the controversy is whether the extinctions resulted dominantly from the actions of man the hunter or whether they were caused by the sudden environmental changes, which it is known, occurred around 11 000 B.P.

There is a good deal of weighty evidence to support the anthropogenic hypothesis (Krantz, 1970). First, outside continental Africa and South-East Asia, massive extinction is unknown before the earliest arrival of prehistoric

Table 4.2

The dates of the major Pleistocene and Holocene mammalian extinctions

Location	Date (B.P.)
North America	11 000
South America	10 000
Northern Eurasia	13 000–11 000
Australia	13 000
West Indies	Mid-post-glacial
Madagascar	800
New Zealand	900
Africa and South East Asia	40 000–50 000

(After Martin, 1967)

man. In America, for instance, there is as yet little conclusive evidence that man had arrived from Asia via the Bering Land Bridge before 12 000 to 13 000 B.P., and certainly, if he had, his numbers were small or relatively localized compared to the time of the so-called Clovis hunters (11 000–12 000 B.P.). The extinctions in North America thus seem to coincide in time with the arrival of man in sufficient quantity and with sufficient technological skill in making suitable artefacts (the bifacial Clovis blades) to be able to kill large numbers of animals. Equally, early man and his dog, the dingo, only arrived in Australia at a time of low sea-level in the latter part of the Würm glacial period. In Africa the massive extinction of animals coincides with the final development of Acheulean hunting cultures which were widespread throughout that continent. In Europe, the efficiency of Upper Palaeolithic hunters is attested by such sites as Solutré in France, where a late-Perigordian level is estimated to contain the remains of over 100 000 horses. The restricted orientation of the subsistence economy of these people is evinced by the concentration on animal representation in their art forms to the exclusion of virtually all naturalistic motifs, with the exception of women.

The hunting activities of man would be made more efficient as he developed more sophisticated tools and learnt to use fire for driving game. Moreover, as Darwin noted on his voyage of the *Beagle*, many beasts which have not known man are remarkably tame and stupid in his presence. It would probably have taken many animal species a considerable time to acquire the knowledge to flee or seek concealment at the sight or scent of man. In addition to the direct effects of hunting activities in reducing their numbers, man may have competed with mammals for a particular food or water supply.

Certain objections have been levelled against the climatic change hypothesis and these tend to support the anthropogenic model. First, it has been maintained that changes in climatic zones are generally sufficiently gradual for beasts to be able to shift with the shifting vegetation and climatic zones of their choice. Secondly, the climatic changes associated with the multiple glaciations, interglaciations, pluvials and interpluvials do not seem to have caused the same abrupt degree of elimination.

Nevertheless, the climatic change model of Pleistocene extinctions still has its adherents, and certain major arguments can be advanced in its favour. Guilday (1967), for instance, has written that '. . . in the absence of man much the same pattern of extinction would have occurred' and he proposed that in the Western Hemisphere man 'may have delivered no more than the final *coup de grace* to isolated remnants already doomed by rapid postglacial environmental changes'. In other words the view is being put forward that some of the environmental changes were rapid, and that man's arrival coincided very largely with them.

A second argument that is raised against any simple anthropogenic model is that in some localities boundaries of a natural type, such as high mountain ranges, did not enable beasts to migrate with the gradually changing conditions (or more rapidly changing conditions) resulting from climatic change. The relatively unaltered state of the African fauna, which still has a fairly large number of large mammals, is due, according to this point of view, to the fact that the African biota is not and was not greatly restricted by any insuperable geographic barrier.

Darwin, in his *Origin of Species* (1936 edn., p. 290), believed in a similar manner that 'As the cold came on, and as each southern zone became fitted for the inhabitants of the north, these would take the places of the former inhabitants of the temperate regions. The latter, at the same time, would travel further and further southward, unless they were stopped by barriers, in which case they would perish.' In Europe the great mountain ranges from the Pyrenees to the Carpathians could have constituted such a barrier, as would the Mediterranean Sea.

Another possible way in which the climatic changes could lead to extinctions is through their effects on mammalian mating habits. Animals with inflexible mating habits are often restricted in range by the season in which their young are born. Slaughter (1967) has maintained that animals with gestation periods of several months would be adversely affected by lengthened winter seasons such as were characteristic of the period 11 000–9500 B.P. They would tend to mate in the autumn and the offspring would then be likely to arrive when there was no grass to sustain them. They would then perish. Animals with relatively short gestation periods (usually, but not always, smaller beasts) tend to await clear signals for optimal weather before mating. It is therefore perhaps significant that it was the larger mammals that were most reduced in numbers during the period of Pleistocene overkill.

Another mechanism of extinction worthy of consideration is disease. This would be especially effective in the case of large mammals because of their lower reproduction rates which would make recovery in population numbers much slower. This again ties in with the relatively higher proportion of large mammals which became extinct. It has been suggested that during glacials animals would be split into discrete groups cut off by ice sheets, but that as the ice melted (before 11 000 B.P. in many areas), contacts between groups would once again be opened up and diseases to which immunity might have been lost because of isolation would spread rapidly. An analogous situation is that by which early European explorers introduced virulent and unaccustomed diseases into the Americas with devastating effects on the native population. Thus any time of rapid population movement resulting from a marked environmental change such as that which marked the end of the Pleistocene and the start of the Holocene could lead to an increased disease risk.

The detailed dating of the European megafauna's extinction lends further support to the climatic hypothesis (Reed, 1970). The Eurasiatic boreal mammals, mammoth, woolly rhinoceros, musk ox, and steppe bison, were associated with and adapted to the cold steppe which was the dominant environment in northern Europe during the glacial phases of the Würm–Weichselian glaciation. Each of these forms, especially the mammoth and the steppe bison, had been hunted by man for several tens of thousands of years, yet managed to survive through the Last Glacial. They appear to have disappeared within a space of a few hundred years—woolly mammoth, woolly rhinoceros, steppe bison, together with horse, saiga, and reindeer, were still present in parts of south-western France during the Bölling (13 500–12 500 B.P., see p. 95), but woolly mammoth, woolly rhinoceros, musk ox, steppe bison, giant deer, and the various cave predators were missing from the western European fauna of the Younger Dryas (around 10 800–10 150 B.P.) when the climate and general environment were otherwise broadly similar to those of the Bölling. Thus the disappearance of this group from western Europe can be pinpointed fairly accurately to the warm period of the Allerød, with its restriction and near disappearance of their habitat.

Post-glacial warmth and disjunct floras

In Britain there are some good examples of how the warming of post-glacial times and the associated spread of forest has led to the fragmentation of the distribution of certain cold-loving floral types which covered a much wider area in the Pleistocene and Early Holocene. Two of the most remarkable 'disjunct areas' are the Burren of County Clare and the Teesdale area of the northern Pennines (Seddon, 1971). Although ecologically very different they both possess certain types of flora which are found scarcely anywhere else in the British Isles. One such plant is the Shrubby Cinquefoil (*Potentilla fruticosa*) which is intolerant of a forest cover and is found currently in its most continuous form in central and eastern Siberia. On the unique limestone pavements of the Burren, and on the river banks and gravels of Upper Teesdale, the plant has been able to withstand the post-glacial warming trends and afforestation.

The effects of post-glacial warmth in creating relict clumps of certain plants in isolated areas is well displayed by the dwarf birch (*Betula nana*). It has been recorded at sites in many parts of Britain in late-glacial and post-glacial deposits, but is now found only in Upper Teesdale and on Scottish mountains. Equally, in north-western Europe there are similar relict areas in the French Jura, in the Harz mountains, and on peat from Lüneburg Heath. It seems clear that this plant, which is from the Arctic-alpine group and is currently found in high-latitudes and at high-altitudes in the Alps, was once widespread over the lowlands of Late Glacial, north-western Europe, but that it has been removed from all localities except on mountains and on some very special habitats with distinctive ecological and microclimatic conditions. This removal has been caused by the spread of forest trees into environments which were formerly suitable.

Man and the classic sequence of Holocene climatic change

Although we have so far outlined some of the ways in which the marked environmental changes at the transition from the Late Glacial to post-glacial times greatly affected man, plants, and animals, it becomes increasingly clear in the Holocene that man himself was an increasingly potent agent of environmental change (Pennington, 1969). For a considerable time it was believed that Palaeolithic and Mesolithic man was relatively ineffectual either because of his small numbers at this stage in his development, or because he did not possess the technological wherewithal. The characteristic Palaeolithic hand-axe, for example, was apparently either a weapon or grubbing tool, and not until the development of the polished stone axe was man equipped with a tool to attack the forest cover of Europe and elsewhere (Smith, 1970).

However, pre-neolithic man did possess the so-called tranchet axe, which could have been effective in forest clearance, but more importantly Mesolithic man may have utilized fire deliberately for driving game and clearing woodland. Sparks and West (1972) believe that fire may have been important as an agent of ecological change even earlier: '. . . the regularity with which hearths are found associated with the Middle and Upper Palaeolithic sites leaves little doubt that Neanderthal Man and his successors were capable of fire production.'

In the British Isles a marked abundance of hazel appears in Mesolithic time and there is no doubt that the European *Corylus avellana* is fire resistant. Of particular interest is the decline of the linden *Tilia*, the pollen of which tends to disappear in many British sites at the same time as charcoal and other evidence of human activity appears (Turner, 1962). This means that the classic VII/VIII zone boundary between the Atlantic and the Sub-Boreal as originally defined by changes in tree-pollen frequencies may have little or no climatic significance. In Switzerland, the first marked fall of the beech curve in the pollen sequence is synchronous with the oldest agricultures in that country (Older Cortaillod culture), and the decline of elm in Denmark coincides with the arrival of the Younger Ertebolle culture (Troels-Smith, 1956). In many parts of Europe the decline of elm around 5000 B.P. may be explained by the use of elm leaves for feeding stock. The leaves were collected for stalled animals. This phase was followed by more intensive clearances for shifting agriculture—the 'Landnam' clearances.

The vegetation and climatic conditions of Holocene Britain

In spite of the role of man in altering vegetational characteristics in the European Holocene, the role of climatic changes in causing some of the observed patterns is considerable.

After the sub-Arctic conditions of the Pre-Boreal (Table 4.1), the Boreal period saw a marked increase in warmth, and conditions were probably relatively dry and continental compared to those of today. The immigrants of the early Boreal included hazel (*Corylus avellana*), and this probably created a scrub beneath a canopy of pine and birch, or else, in some localities, a pure hazel woodland. In the Late Boreal more warmth-loving plants such as elm (*Ulmus*) and oak (*Quercus*) appeared in greater numbers, so that the Late Boreal appears to have been the last time at which pines grew generally in England on soils of all types. Subsequent to this time in England, Wales, and Ireland pine seems only to have grown locally, probably on the poorer soils on which it is familiar today. At the transition between the Boreal and the Atlantic elm and oak spread further, and warmth-loving plants such as the Lime (*Tilia*) appear. Rather dry conditions, which had led to the reworking of deposits marginal to lakes and to the drying out of some mire surfaces, were replaced at the very end of the Boreal by wetter conditions conducive to the growth of wet, peat-forming communities of *Eriophorum* and *Sphagnum*. In the Atlantic itself, when warm, wet conditions prevailed in Britain, the extreme oceanity of climate brought about replacement of the deciduous forest on flatter surfaces over about 360 m in altitude. In their place blanket-peat-forming communities developed, though on steeper slopes with well-drained soils deciduous forests were able to extend to at least 760 m, and in Ireland the forest extended over great lowland areas now covered by blanket-peat and raised bogs. At this period grassland was rare except above around 900 m, and open habitats must have been greatly restricted to such specialized niches as unstable scree, limestone pavement, and coastal shingle, sand, and silt. In addition, over much of England pine practically disappeared, and the forest consisted of oak, elm, alder, and lime, with birch in the north and west but little in the south and east. This was the period of the widest spread of *Tilia*, the most exacting in its climatic requirements of the British forest trees. Most forests were of the Quercetum mixtum type, in places pure oakland, but in places a more complex deciduous mosaic with elm and lime. Pine was restricted to the Scottish highlands.

Changes in post-Atlantic vegetation owe relatively less to climate than the changes of the Boreal and the Atlantic, and man and soil deterioration assume great importance. The role of man as an agent in creating the boundary between the Atlantic and the Sub-Boreal vegetation types has already been commented upon. The role of soil deterioration is less easy to assess, though intense leaching of glacial sediments under the warm, wet conditions of the Atlantic may have led to the development of podzols and other soils relatively inimical to the deciduous forest (Pearsall, 1964). Hardpans of the podzolic type may also have led to some waterlogging of soils by impeding their drainage, and they would also have been relatively acid in type.

It is conceivable that new agricultural techniques could, from Neolithic times onward, have accelerated this podzolic condition (Mitchell, 1972).

Podzolic soils would in turn encourage blanket bog formation. Burning and ploughing would help to release minerals which would accumulate as hardpan. This hardpan by impeding drainage would give ideal conditions for blanket-peats to accumulate, and in Ireland and western Wales neolithic fields, occupation sites, and megalithic tombs are sometimes found buried by peat. However, peat bogs or blanket mires, the development of which coincides with the elm decline (5300–5100 B.P.), do not always occur above well-developed podzol horizons (Moore, 1975). Indeed, sub-peat profiles are often immature, suggesting that the soil is not always the predominant cause of mire development. Another possible factor in the timing of peat-bog development was the removal of the natural tree canopy by mesolithic or neolithic man. This would reduce the transpiration demand of the vegetation and would also reduce the degree of interception of precipitation. Hence more water would be available to raise groundwater and soil-water levels, thereby favouring peat development. In the southern Pennines the basal layers of all marginal peats (Tallis, 1975) contain widespread evidence of vegetation clearance by burning, either in the form of microscopic carbon particles (similar to contemporary 'soot'), small charred plant fragments, or larger lumps of charcoal.

Thus peat bogs, in that their development could be promoted by climatic change, soil maturation, and by human interference in the highland ecosystem, illustrate the complexity of factors that may be involved in any environmental change.

The American Holocene sequence
It is interesting to compare the American sequence with that of Europe and Britain. Although a simple cold-hypsithermal-cool sequence has found some favour, it now seems that the sequence was at least as complex as that in Europe. The following sequence has recently been put forward for the Northern Great Plains (Table 4.3). As can be seen it utilizes the European terminology in large part. Further north, in Canada, there have been comparable attempts (Table 4.4) to correlate the American sequence with the European, and it has been suggested that, as human interference has been less in the Canadian Holocene, that the Canadian sequence gives a more realistic impression of the role of climatic change in the development of post-glacial vegetation associations, and that by using the Canadian sequence as a standard one can assess the importance of man as opposed to climate in certain major vegetational changes such as the *Ulmus* decline. In central Canada there appear to have been changes that are broadly comparable to those in Europe: the extension of the forest between 6500 and 5000 B.P. for example correlates with part of the classic Atlantic in Europe, while the retreat of forest after 2500 B.P. appears to correlate with the cold, wet, and oceanic conditions of the European Sub-Atlantic.

Table 4.3

Holocene environmental changes of the Central Great Plains, U.S.A.

Full Glacial	to 13 000 B.P.
Late Glacial	13 000 to 10 500
(with minor advances such as the Valders and Two Creeks). Spruce over much of Northern Plains	
Pre-Boreal	10 500–9 140
Boreal	9 140–8 450
Atlantic	8 450–4 680
(maximum penetration eastwards of grassland *c*.7 000 B.P., and expansion of species of the coniferous and deciduous forests northwards of their present limits)	
Sub-Boreal	4 680–2 690
(some cooling, and grassland retreats more or less to present position)	
Sub-Atlantic	2 890–1 690
(possibly wetter)	
Scandic	1 690–1 100
(return towards conditions of early Atlantic time—drier)	
Neo-Atlantic	1 000–760
(warmer climate continued, but probably wetter)	
Pacific	760–410
(Around A.D. 1200 a shift to drier conditions is noted)	
Neo-Boreal	410–115
(colder, moister conditions)	
Recent	115–
(increase in strength of dry westerlies. Warmer, drier)	

(After Hoffmann and J. Knox Jones, 1970)

Table 4.4

Holocene environmental changes in Central Canada and NW. Europe

Central Canada	Yrs. B.P.	North-western Europe
Forest retreat, expansion of tundra, peat growth ceases at Ennadai Lake		Recurrence surfaces, Greenland colonists perish, Little Ice Age
	700	
Small northward extension of forest		Retardation layers in peat. Exploration of north Atlantic
	1500	
Retreat of forest to south of Ennadai		Recurrence surfaces, Alpine glaciers advance
	2500	
Alternations of cool and warm climate		Alternations of cool and warm climate, recurrence surfaces and retardation layers in peat
	3500	
Small retreat of forest		*Ulmus* decline
	5000	
Forest extended far north		Continuation of climatic optimum
	6500	
Rapid deglaciation, swift immigration of forest		Beginning of climatic optimum. Warmest period of Post-Glacial
	8000	

(After Nichols, 1967)

The Holocene in East Africa

In the last decade there have been a number of studies, utilizing pollen-analysis of climate-vegetational changes in the East African mountains. The pioneer work was done by the South African 'Bloemfontein School' and this seemed to show that it was possible to correlate Late Glacial and Post-Glacial events in the East African mountains with the European sequence. A bore pit down in the Kaisungor Swamp in the Cherangani Mountains gave the succession shown in Table 4.5. A little later a core from the northeast flank of Mount Kenya at Sacred Lake (Coetzee, 1964) indicated that

Table 4.5

Holocene environmental changes in the East African mountains

Date B.P.	Vegetation	Climate at 2926 m	European equivalent
	Tree-line descending	Wet, becoming colder	Sub-Atlantic
2 650			
	Forest maxima	Warmer Colder	Sub-Boreal
4 960			
	Forest closed, tree-line rising	Warmer and wet. Temperature rising	Atlantic
7 740			
	Tree-line round swamp	Warmer and relatively wetter	Boreal
8 625			
	Open vegetation, Tree-line below swamp	Cold and dry	Younger Dryas
9 990			
	Compositae maximum	Warmer and dry	Allerød Older Dryas
10 790			
	Maxima of grasses and ericaceous belt	Cold and dry	Bølling
	Alpine vegetation	Very cold and dry	

(After Van Zinderen Bakker, 1962)

montane forest began to show a definite development after 10 583 B.P. (coeval with the end of the Late Glacial and beginning of the European Pre-Boreal) and to replace an open grassland and heath vegetation. Similarly, a core from Muchoya Swamp in Uganda (2256 m) indicated a change from Montane heath and grassland to *Hagenia* Forest around 11 000 B.P. This implies a change from a cold, dry climate to moister and warmer conditions (Morrison, 1968).

However, the type of intercontinental correlation revealed in Table 4.5 has been severely criticized by Livingstone (1967), who has stated that 'There is no discernible basis for a detailed correlation of either vegetation or climate on a zone-for-zone basis with the established sequences of temperate countries'. Nevertheless, the marked pollen changes established both by the South African and other workers does mean that in the Holocene the mountainous regions of East Africa have been as unstable with respect to vegetation—and probably climate and fauna as well—as extra-tropical regions. As in Europe, forest vegetation seems to have been more extensive for part of the Holocene than it is now, with the maximum being between about 5000 B.P. and 2650 B.P., and to have retreated somewhat after this time.

Post-glacial times in the Sahara and adjacent regions

In the Sahara desert it has for long been suspected that climatic conditions were wetter at some stage or stages in the Holocene than they are at present. This was deduced from facts such as the widespread distribution of rock paintings, and of human stone and other tools, in areas which are currently far removed from waterholes. Certain of the species represented in rock painting, notably elephant, rhino, hippo, and giraffe were regarded as being representative of a moderately to strongly luxuriant savannah flora. Pollen-analysis, though so far on a limited scale and subject to many doubts, has confirmed this essentially subjective archaeological evidence, and pollen of Aleppo pine and other trees have been found in sediments of Holocene age in the Hoggar and other massifs. There are also now a large number of radiocarbon dates for lacustrine sediments in various parts of the desert which enable one to establish the sequence of events with a little more certainty than hitherto. It seems likely on the basis of dates from Chad, Ténéré, the Nile valley, the Saoura valley, and the Hoggar that there were three lacustral phases in the Early Holocene (before about 8500 B.P., from 7050–4150 B.P. and from 3550–2450 B.P.), and that during them vegetation was denser than at present.

In the desert to the west of the Nile there are numerous tree stumps of acacia, tamarisk, and also of sycamore (*Ficus sycomorus*). It is significant that these stumps, with diameters of 30–40 cm and a density of 5–11 per hectare indicate that an open savannah existed in this sub-pluvial some 200 km farther north than this vegetation can survive today (Butzer, 1961).

At the Kharga Oasis there are immense deposits of lime-rich spring tufas around or in which neolithic tools have been found in great numbers. This indicates higher groundwater levels and a considerable population. The Neolithic was a time particularly favourable for human activities in the Sahara (Faure, 1966).

An indication of the significance of the 'neolithic pluvial' in comparison with other climatic phases which we have discussed is given in Fig. 4.3. This illustrates the boundaries of major vegetation belts in the Sudan as reconstructed by Wickens (1975) on the basis of palaeobiological evidence. The present-day boundaries (B), of desert, semi-shrub desert, grassland, and lowland forest, are to the south of those of the wet period 6000–3000 B.P. (D). On the other hand the very wet Early Holocene vegetation boundaries (C) are considerably further to the north than those of the 'neolithic pluvial'. Over-all it appears that during this sub-humid period the climatic and vegetation belts were about 250 km north of their present position.

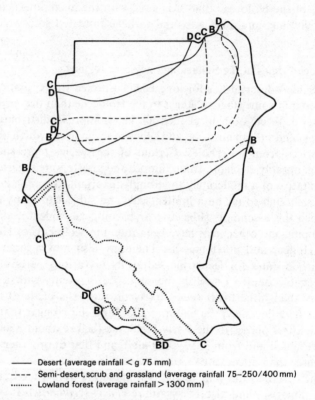

——— Desert (average rainfall < g 75 mm)
- - - - Semi-desert, scrub and grassland (average rainfall 75–250/400 mm)
········· Lowland forest (average rainfall > 1300 mm)

FIG. 4.3. Vegetational zone changes in the Sudan in the Late Pleistocene and Holocene.
A = Dry period 20 000–15 000 B.P.
B = Present day.
C = Very wet period 12 000–7000 B.P.
D = Wet period 6000–3000 B.P.
(from data in Wickens, 1975, Figs. 2, 3, 4, 5).

The Holocene in northern India

Apart from this work undertaken in East Africa there are relatively few studies of the pollen record in tropical lands, though in India some recent studies by Singh (1971) have enabled one to assess the nature of Holocene changes in one of the most important areas of early world civilization—the

Indus Valley and northern India. Singh's ideas have shown clearly that some of the earlier ideas on the progressive post-glacial desiccation of Asia are no longer tenable in Rajasthan, and that, as elsewhere in the world, the climate of post-glacial times has been both drier and wetter than that at the present. His work is based on detailed pollen-analysis and stratigraphic examination of sediments from lake basins, such as Sambhar (27 °N, 75 °E), and Lunkaransar (28° 30 °N, 73° 45 °E) in Rajasthan.

The first phase of the Holocene is represented by a thick layer of aeolian dune sand beneath the lakes. This layer grades into the numerous fossil dunes of the eastern parts of Rajasthan, most of which seem to be earlier in age than the arrival of Mesolithic man in profusion over the arid zone. Two radiocarbon dates, which seem to correspond well with dates from some parts of Africa (see p. 82), indicate that this arid phase was succeeded by deposition of lacustrine deposits around 9000–10 000 yrs B.P., dates which also tie in with the archaeological evidence that man was able to enter the area in large numbers after the humid conditions which led to dune stabilization had arrived (Singh, 1971). The lake sediments (9250 ± 130 B.P. and 9260 ± 115 B.P.) are of an essentially freshwater type and they consist of laminated clays containing the pollen of sedges and grasses, but not any marked degree of halophytes. Artemisia, which now grows under conditions with 500 mm or more of precipitation per annum, notable in the foothills of the Himalayas, was abundant, and judging from the absence of sand intercalations the vegetation had been successful in stabilizing the dunes. In Phase III (9500–5000 B.P.) the rainfall appears to have been slightly lowered, and halophytic vegetation, represented by Chenopodiaceae/Amaranthaceae, started a gradual rise. This phase also witnessed scrub burning by early man, as carbonized vegetable remains are present in the sediments. In Phase IV wetter conditions returned between 4950 and 3750 B.P. Thereafter the climate shows a small-scale oscillation to drier conditions between 3450 and 3750 B.P. with signs of drying up in the lakes.

This drier sub-phase coincides chronologically with the decline of the Indus (Harappan) civilization which saw the growth of the great settlements of Mohenjo-Daro, Harappa, and Kalibangan. Some archaeologists believe that the sheer size and extent of cities and other settlements of the Harappan in the Indian Desert is evidence in itself for somewhat moister conditions during the period when they flourished. It has been estimated, for example, that Mohenjo-Daro was inhabited by up to 40 000 persons and had an outer circuit of some 5 km. It was probably at its peak around 4000 B.P., during Singh's Phase IV of wetter conditions. Other archaeologists have used indirect evidence provided by drawings and engravings of animals and plants on potsherds, by animal and wood remains, by flooding horizons, and by extensive 'gabar bands' (dam-like structures constructed artificially for water conservation by prehistoric man) to support their theories on formerly higher precipitation in Harappan times. Much of the evidence is questionable in itself, but taken as a mass together with the pollen data it may have some validity. Whether climatic deterioration was temporally *and* causally related to the decline of the Harappan civilization is a further problem, and the role of massive floods, Aryan invasions, or soil deterioration through over-use, have to be borne in mind in addition to the climatic factor.

However, some of the most noted results of environmental change in the dry zone of northern Indian stem not from simple climatic change, but from massive shifts in the courses of some of the great rivers draining from the Himalayas. The relief of the interfluve between the two main systems, the Indus in the west and the Ganga (Ganges) in the east, is slight, and so it is relatively easy for river capture to lead to the diversion of river water from one great system to the other. This process led to the loss of discharge from rivers over a considerable portion of the Punjab in the Holocene. About 4000 years ago the present waters of the great Yamuna (Jumna) flowed towards the Indus, and to the Arabian Sea. This was achieved by means of a river called the Chautang which flowed through the locations of the present cities of Hissar and Saratgarh. Then between 1500 and 900 years ago the Yamuna captured the headwaters of the Chautang and as a consequence its flow was diverted to the other great river system, the Ganga. The palaeochannels of the major abandoned river system still occupy the alluvial plain between the Yamuna and the Sutlej rivers.

The post-glacial climatic optimum and Neo-glaciation

Whatever may have been the sequence of environmental changes which took place in different areas during the course of the Holocene, one of the most contentious but interesting problems is that of the so-called climatic optimum (Manley, 1966).

That climatic conditions were appreciably warmer during a section of the Holocene than they are now, was discovered by Praeger, who detected it in connection with his investigation of the fauna of the estuarine clays of the north of Ireland. This was in 1892. From a comparison of the fauna of the Belfast estuarine clays with that of the present shores of Ireland, he gave the first proof of a definitely higher temperature during their deposition. Scandinavian workers subsequently confirmed this finding by an examination of the fauna of the *Tapes* Submergence (see p. 195) in the Oslofjord. They also discovered that in Lapland the pineforest had at some stage in the Post-Glacial moved into zones which are now dominated by birch or alpine associations. This period of extended distribution has been called the post-glacial climatic optimum. The tree-line extended 500 m higher than today in northern Europe, and the treeless tundra almost disappeared from northern Siberia. In Norway where the maritime influence was stronger the tree-line displacement was less—only 300 m. One of the most important markers of this optimum was the European land tortoise (*Emys orbicularis*) which spread into Denmark at this time, but disappeared in the Sub-Atlantic. Cool and damp summers are highly unsuitable for the animal, especially to the development of its eggs (Godwin, 1956). Another important indicator of the post-glacial warm period is the hazelnut (*Corylus avellana*). Its present distribution in Scandinavia, shown in Fig. 4.4 is markedly different from its distribution 5000–6000 years ago, for at that time it extended further north, and higher in altitude than at the present time.

In Greenland too there is evidence that at some point during the Holocene conditions were more favourable to life than they are now. The

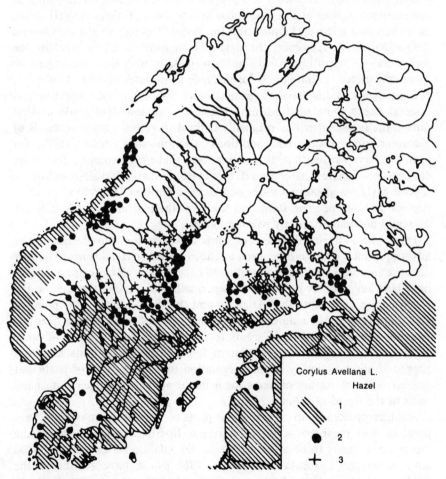

Fig. 4.4. Indications of the post-glacial warm period illustrated by the Present and post-glacial distribution of the hazelnut (*Corylus avellana* in Scandinavia).
(1) = Present general distribution.
(2) = Current records of individual occurrences.
(3) = Hazelnut fossils in sediments of the post-glacial warm period.
(From Frenzel, 1973, Fig. 2.)

edible mussel, *Mytilus edulis*, which now has a northern limit in Greenland waters at about 66 °N, is found in raised beaches, which have been dated at about 5000–7000 B.P., which occur at a considerably higher latitude—73 °N. Another 'southern' bivalve, *Chlamys islandica*, is also found in areas outside its present range (Funder, 1972).

However, some dissatisfaction has been expressed from time to time with the use of the word 'optimum', especially as in rather drier areas the increased temperature would be far from advantageous for plant growth. Various other terms have therefore been introduced including 'altithermal' and 'hypsithermal'. The latter was proposed in 1957 by Deevey and Flint as a term to cover four of the traditional pollen zones (V through VIII in the Blytt–Sernander System) embracing the Boreal through to the Sub-Boreal (8950–2550 B.P.). However, the dates which other workers give for the 'optimum' or 'hypsithermal' do not always tally with this, tending to be more constricted. Lamb (1969) for instance gives dates of 6950–4350 B.P.

Subsequent studies have suggested that a concept of one hypsithermal interval needs some modification. There is now abundant evidence that during the hypsithermal as originally defined there were some renewals of glaciation and some cold conditions. Denton and Porter (1970), for example, have written that 'It is now known that rather complex low-order changes of climate characterized the hypsithermal interval, resulting in several early neoglacial episodes of glacier expansion. Therefore, in some regions at least, neoglaciation and the hypsithermal interval, as they are currently understood, partly overlap in time.'

Figure 4.5 indicates the dates of some of the major 'neo-glacial' glacier advances which seem to have been a feature of post-glacial times. The new data are important in that conventionally the Alpine glaciers, for example, have been regarded as having remained smaller than today throughout the warm period embracing the 'Atlantic' and the 'Sub-Boreal'. This point of view can no longer be maintained, and the period of transition between the Atlantic and the Sub-Boreal seems to have been one of glacial advance (5200–4600 B.P.). This interruption of hypsithermal conditions has been suggested additionally by certain vegetation indicators. Also of particular interest is the clustering of dates for a major advance of glaciers in many parts of the world at around 2800 B.P.

Another problem is that in different parts of Europe the maximum temperature was recorded at different times. In Sweden, for instance, the 'optimum' appears to be associated with the Atlantic period (around 6000 B.P.), when the Littorina Sea (see p. 196) would have reinforced the mildness of the winters. In Denmark, on the other hand, the optimum appears to have been around 4000–3000 B.P. (the Sub-Boreal). In Scotland too, trees reached their highest limit on the Cairngorms in the Sub-Boreal (Manley, 1966).

However, although the nature and timing of the optimum may be the

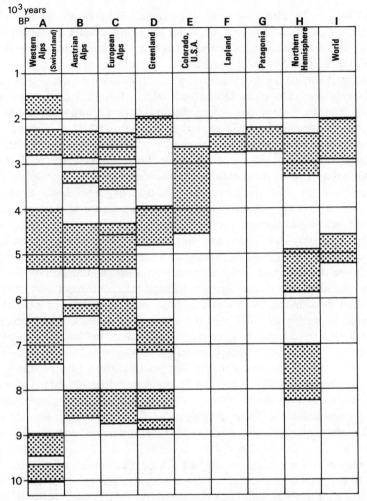

F{\scriptsize IG}. 4.5. Radiometrically dated neo-glacial advances (shaded) of the period 10 000 to 1000 years ago for miscellaneous parts of the world. From data in Patzelt, 1974, (A); Bortenschlager and Patzelt, 1969, and Patzelt, 1974, (B); Patzelt, 1974, (C); Brink and Weidick, 1974, (D); Benedict, 1968, (E); Karlén, 1973, (F); Mercer, 1969, (G); Denton and Karlén, 1973, (H); Bray, 1974, (I).

subject of dispute, it does appear from many sources that temperatures have been appreciably higher during certain parts of the Holocene than they are at present, though the degree of temperature amelioration is rather variable. A positive deviation over the present mean of from 1–3 °C is suggested for many areas in the temperate zone.

The hypsithermal has also been recorded through oxygen-isotope measurements of an ice core at Camp Century on the Greenland Ice Cap. This has indicated that a warm phase lasted from around 4100 to 8000 B.P. with extremes close to 5000 and 6000 B.P. (Dansgaard *et al.*, 1970). It was replaced by colder conditions between 2100 and 2500 B.P.: dates which correspond fairly closely to a neo-glacial glacial advance, and the early part of the classic, Sub-Atlantic deterioration.

Among effects of the hypsithermal phase, the following are probably the most important: the Arctic Ocean was largely free of ice, at least in the summer months; both Lake Bonneville and Lake Lahontan in the U.S.A. dried up completely; the Caspian *may* have shrunk to 50 m below present sea-level which is 22 m below its present position; and, in the Sahara and parts of the near east, there was a moister climate than now. This has been already referred to as the 'neolithic moist interval'.

It seems likely that the increased humidity of some desert zones was associated with a northward expansion of the equatorial rainbelt and its fringe of occasional summer monsoon rains, whilst the warm conditions in high latitudes may have been the result, says Lamb, of the displacement of the main subpolar depression tracks, and the axis of the subtropical high pressure belts by around 10° of latitude over Europe (Lamb, 1969).

The spread of early man through Europe may also have been partially influenced by climatic factors at this time. A major phase of agricultural colonization in Europe went forward under mild Atlantic conditions from the sixth to the middle of the second millennium B.C. and largely involved the cultivation of wheat. The Campaignien peoples entered from North Africa, and the Danubian peoples from the steppes of eastern Europe during this phase (Demougeot, 1965). Subsequent climatic deterioration led to the decline of wheat cultivation, which in many areas was supplanted by oats and rye. Indeed one oat, *Avena fatua*, was particularly favoured by the increasing length of cold winters for its seed is not viable if winters are not cold (Isaac, 1970).

A related factor which greatly influenced the spread of the neolithic agricultural revolution through Europe was the great belt of wind-blown loess (see p. 43) which had been deposited extensively over the plainlands. It was at one time thought that these soils, because of their relatively dry soil climate when compared to, for instance, the damp, heavy, clay soils developed on the boulder clays of the North German Plain, might have supported a much more open heath-like steppe vegetation which would have favoured early man. The clays on the other hand would tend to support a denser forest which it would be relatively difficult to clear for agricultural purposes. However, more recent pollen-analysis of deposits in the loess terrains do not fully substantiate this point of view, for they indicate that the loess too was covered by a forest.

For the same reasons it has become necessary to lay aside another early theory which tried to relate a dry or 'xerothermic' phase during the classic Sub-Boreal of the Blytt–Sernander model with a decline in forest cover. Under this theory, Gradmann's 'Steppenheidetheorie', it was held that drought conditions were such that forest could not be maintained on the lighter loessic soils of central Europe, so that there existed a corridor of open steppe along which the fauna, flora, and prehistoric peoples of south-eastern Europe migrated into the oceanic west.

Nevertheless, the correspondence between settlements of Danubian type and the loess soils is remarkable. The explanation for this is probably that whilst these soils were not related to open vegetation conditions except locally, they were well-drained, fertile, and easy to till, whereas other types of soil would in general have been heavy, cold, and ill-drained. Moreover, recent investigations suggest that neolithic man was much more competent technologically to clear woodland than had previously been believed. Thus woodland may not have been quite the barrier envisaged at first sight.

In the Americas there has been a considerable amount of discussion as to the effects of the Holocene climatic fluctuations, for although man came late to the Americas, once he became established, he built up quite large populations. The south-west dry zone of the United States, for example, was extensively populated between 14 000 and 10 000 B.P. However, relatively few prehistoric sites are encountered in that area between 10 000 and 4500 B.P. (Griffin, 1967), and some archaeologists take the probably extreme view that the Prairies and western plains were almost abandoned between 6000 and 4500 B.P., only to be recolonized after that time. The period of greatest population sparseness appears to coincide with the allegedly dry Altithermal (Stephenson, 1965; Irwin-Williams and Haynes, 1970), while periods of demographic advance (such as the Folsom occupation of 10 800–10 300 B.P. and the post-4500 resurgence of population) coincide with improved moisture conditions. Although there is some doubt as to how dry the Altithermal was (see, for example, Martin, 1963), it is widely regarded as being a major factor in the prehistory of the drier parts of America. As Malde (1964, p. 127) has written, 'As the Altithermal drew to a close, relatively wetter and colder conditions returned, and the tempo of human life accelerated, along with evident growth in population.'

The little optimum, A.D. 750–1300

After the climatic optimum of the Middle Holocene, conditions once again became cooler in many regions, but in early medieval times there was a return to more favourable conditions—the so-called 'little optimum'.

From about A.D. 750 to 1200–1300 there was a period of marked glacial retreat which on the whole appears to have been slightly more marked than has been that of the twentieth century. The trees of this phase, which were eventually destroyed by the cold and glacial advances from around A.D. 1200 onwards, grew on sites where, in our own time, trees have not had time, or the necessary conditions, to grow again. In terms of a more precise date, the medieval documents that are available place the most clement period of this optimum, with its mild winters and dry summers, at A.D. 1080–1180. At this time, the coast of Iceland was relatively unaffected with ice, compared to later centuries, and settlement, as will be seen, was achieved in now inhospitable parts of Greenland. It is also believed that the relative heat and dryness of the summers, which led to the drying up of some peat bogs, was responsible for the plagues of locusts which in this period spread at times over vast areas, occasionally reaching far to the north. For instance, during the autumn of 1195 they reached as far as Hungary and Austria. In northern Canada, west of Hudson Bay, a fossil forest has been discovered up to 100 km north of the present forest limit, and four radiocarbon dates from different sites show that this forest was living about A.D. 870–1140. It is also interesting that the Camp Century ice core from Greenland has revealed to American and Danish workers that a cold wave is evident after about 1130–1160, but that for five centuries preceding this there was a phase of appreciable warmth. This has also been confirmed by a more recent core at Crête (Central Greenland) (Dansgaard *et al.*, 1975).

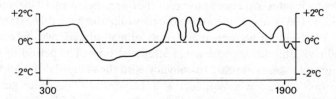

FIG. 4.6. Chinese temperature patterns based on miscellaneous phenomena (appearance of frost, freezing of rivers, blossoming of trees and flowers, migration of birds, etc.), gazetteers, and instrumented observations (after Hsieh, 1976).

One additional line of evidence that has been utilized to gain an appreciation of the nature of this phase is the presence of vineyards in various parts of Britain. Domesday Book (1085) records thirty-eight vineyards in England besides those of the King. The wine was considered almost equal with the French wine in quality and quantity as far north as Gloucestershire and the Ledbury area of Herefordshire, the London Basin, the Medway Valley, and the Isle of Ely. Some vineyards even occurred as far north as York, and Lamb (1966) regards this as being indicative of summer temperatures 1–2 °C higher than today, a general freedom from May frosts, and mostly good Septembers. In China, at about the same time, lychees, sensitive trees which succumb at temperatures below −4 °C, were an economic crop in the Szechuan Basin in western China, but today they are limited to the south of Nanling. Miscellaneous evidence of this type was used to construct Fig. 4.6, the pattern of which suggests a striking correspondence with the fluctuations derived from the Greenland ice core (Hsieh, 1976).

The little optimum and agriculture in North America

Just as the main climatic optimum seems to have affected people in the south-west of the United States, so the little optimum coincided with a radical change in the fortunes of the agricultural peoples of the Missouri Valley, Dakota, and the South West.

Agricultural societies in the Missouri and Upper Mississippi valley and the Great Lakes–north-eastern U.S.A. had their primary growth and development from A.D. 700 to 1200, and this has been related to the climatically favourable neo-Atlantic episode which brought moist tropical air over the great plains (Malde, 1964). This favoured both corn growing and game. Agriculture was practised over its greatest extent, the material manifestations of cultural vigour achieved notable local diversity, and in the Arctic, the Thule coastal whaling culture had a remarkable development and spread. This phase seems to coincide with the little climatic optimum of Europe (Griffin, 1967).

Shortly before the thirteenth century a very rapid withdrawal of agricultural peoples took place both in the Missouri Valley of North and South Dakota, and in the south-west (Leopold et al., 1963). In the latter, within less than three centuries, the area occupied by farmers shrank from 600 000 square km to about 220 000 square km (Woodbury, 1961). This phase, during which population became much concentrated into favourable areas, coincides with the climatic phase called Pacific I (A.D. 1200–1300) in which, some authorities maintain, an increased flow of westerlies across the northern plains led to the introduction of greater amounts of cool, dry air (Lehmer, 1970). Both the alleged dryness, which necessitated concentration at favourable points for irrigation, and coldness, which greatly reduced the growing season and the altitudinal range for corn cultivation, encouraged this contraction.

Another probable cause of this marked population change was the extensive erosive gullying of alluvial fill upon which the main agricultural societies depended. The pueblo people cultivated through making use of floods over wide valley floors. Thus their prosperity was favoured by alluviation and hampered by erosion. Gully erosion would also lower local water-table levels. Bryan believed that these erosive phases were caused by a swing to drier conditions as revealed in tree-ring analysis (Bryan, 1941). This period was also the period of striking expression of religious ceremonialism intended to influence or produce rain for agricultural fertility. In some parts there is evidence for a shift from dependence on agriculture to hunting of beasts like the hapless bison.

It has to be pointed out, however, that the role of aridity in promoting changes of this type is not completely proven. In parts of the south-west pollen-analysis, in apparent contrast to some of the tree-ring evidence, suggests conditions were if anything somewhat wetter. Pine pollen increased relative to pollen of Compositae and Chenopods. Moreover, some of the arroyo cutting (gully erosion), judging from an analogous situation at the end of the nineteenth century (see p. 163), may have resulted not from drought reducing the vegetation cover but from an increased incidence of heavy summer storms. It also needs to be stated that some archaeologists have shown that harassment by nomadic raiders, the Athapaskans, who were ancestral to the present Navajo and Apache, could explain the observed tendency towards both concentrated settlement and the abandonment of certain sites (Jett, 1964).

Conditions improved after 1450 (Pacific II phase lasting until A.D. 1550) and there was once again a much more extensive occupation of the Missouri Valley in South Dakota by the village tribes.

The last Little Ice Age (neo-glaciation)

One of the most significant Holocene environmental changes, not least because of its effects on the economies of highland and marginal areas in Europe, was the renewed phase of glacial advance since the late medieval period. This phase has often been called the Little Ice Age, but recently the term 'neo-glaciation' has been proposed 'to encompass the interval of rebirth or renewed growth, and all subsequent fluctuation, of glaciers after the time of maximum hypsithermal glacier shrinkage' (Denton & Porter, 1970, p. 102). Its effects have been noted all over the world. In China, for example it was at its peak from 1650–1700 (Chu Ku-Chen, 1973), but the date at which the late-medieval Little Ice Age began is variable from area to area. The Great Aletsch Glacier in the Swiss Alps advanced over part of an aqueduct used to transport meltwater to a local village as early as the thirteenth century. Similarly, the Chickamin Glacier in the Cascade Range of Washington State (U.S.A.) reached its maximum in the thirteenth century. In the S. Tyrol there was a major advance A.D. 1150–1250 (Mayr, 1964). In most areas, however, the maxima were reached at various times from the middle of the fourteenth century to the middle of the nineteenth century.

In Norway, where the glacial advances are relatively well chronicled by tax records and other sources, the advances appear to have begun between 1660 and 1700. The first half of the eighteenth century was marked by a general advance which amounted to several kilometres for some glaciers and culminated between 1740 and 1750. After this there was some recession, interrupted by re-advances, notably in 1807–12, 1835–55; 1904–5, and 1921–5. These have tended to leave small moraines. These later advances did not usually manage to reach the position gained in 1750.

An examination of land rent assessments from the Jostedalsbre region of western Norway, and of documents concerned with applications for their reduction has provided J. M. Grove (1972) with detailed information about the incidence of landslides, rockfalls, floods, and avalanches during the

Little Ice Age in Norway (Fig. 4.7). The evidence makes it clear that there was a much increased incidence of major mass rock movements and floods in the late seventeenth century and on until the nineteenth century. Moreover, this environmental change began abruptly with a marked clustering of disastrous incidents between 1650 and 1760, and in certain years during

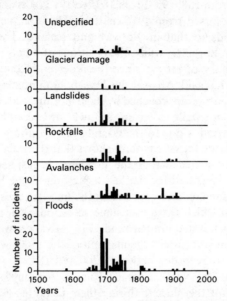

FIG. 4.7. The incidence of Little Ice Age mass movements in the Norwegian parishes of Oppstryn, Medstryn, Løen, and Olden as revealed by Landskyld (land rent) records (after Grove, 1972, Fig. 2).

that period, such as 1687, 1693, and 1702. Conditions for farming were thus very much less favourable than they had been in previous centuries.

The Icelandic glaciers and ice caps show a broadly similar history during this period. From the time of the first colonization of the island around A.D. 900 until at least the fourteenth century, the glaciers were considerably less extensive than they were after about 1700. There was a general advance in the early eighteenth century which reached its maximum around 1750. From 1750 to 1790 the ice tended to be relatively stagnant or to be in a state of retreat, but it advanced again in the early nineteenth century and in some cases had, by 1840–60 reached a more forward position than the previous maxima of 1750. A general recession towards the present position took place after about 1890.

In the Alps the situation is extremely well documented compared to other areas. From about 1580 onwards tracts of cultivated land and forests were covered by advancing ice, and the local people were also subjected to greater flood risk. The local economies suffered and a series of supplications for tax relief were made. The Rhone Glacier advanced strongly from 1600 to 1680. From the mid-seventeenth to the mid-eighteenth centuries the glaciers were relatively quiescent, though still at a more forward position than in 1600. From the mid-eighteenth century there was a major phase of advance, divided into three main stages: 1770–80, 1818–20, and 1835–55. Retreat was then fairly general between 1850 and 1880, only to be replaced by some advance from 1880–95. From 1895 to 1915 recession continued, but was temporarily reversed from 1915 to 1925. Thus the picture in the Alps broadly corresponds to that in Norway and Iceland. The various stages are often visible as moraines, whilst the positions of hotels and other tourist facilities tell the story of retreat since the nineteenth-century maxima. As Jean Grove (1966) has said 'All over the Alps mountaineering huts stand high above the ice and must be approached by steep moraines, fixed ladders or even ropes. They were not built to ensure an awkward scramble at the end of the day; their isolation is due to the wasting of ice during the last century.'

The American Little Ice Age pattern shows that the glacial advances were broadly contemporaneous across the northern hemisphere, although an advance has been suggested for the Sierra Nevada about 1000 B.P. (Curry, 1969). The maximum advance occurred in the middle 1600s and lasted until around 1700, from which there was some recession, then an advance, and then recession, which lasted until around 1785. Advances were characteristic of the 1800s, and in some areas the maxima of the 1700s were exceeded. In Alaska maxima were recorded between 1700 and 1835.

Some data from the North Cascade Range, Washington, allow one to compare the state of the glaciers during their Little Ice Age maxima with their state at the present. One can see that in general their termini are now on average around 300 to 400 m higher in terms of altitude, and that their areas have been reduced by 50 to 60 per cent (Table 4.6).

Table 4.6
Glacial fluctuations in the Dome Peak area, Washington State, U.S.A.

	South Cascade	Glacier Le Conte	Dana	Chickamin
Century of maximum advance	16th/17th	16th	16th	13th
Area of glacier (km)² at maximum	4·15	2·98	3·99	7·80
Area of glacier (km)² in 1963	2·72	1·58	2·46	4·87
Ratio (min./max.)	1 : 1·5	1 : 1·9	1 : 1·6	1 : 1·6
Altitude of terminus at neo-glacial maximum (m)	1490	1340	1270	1100
Altitude of terminus in 1964 (m)	1615	1829	1768	1525

(After Miller, 1969)

The glaciers of Greenland were also affected by the Little Ice Age, and advanced strongly between A.D. 1700 and 1850. The maximum extensions were, however, reached rather later than in many parts of the world. In south-west Greenland the maximum was around 1850. At the inland ice margin the maximum was reached around 1890/1900, whilst in the north-west the maximum was not reached until 1915/25. As elsewhere, however, there was a general retreat along the whole Greenland west coast between 1920/25 and 1940/45. Since the latter date there has been a decelerating rate of retreat.

The Antarctic area seems to have escaped the cold epoch until rather late, a factor which doubtless helped the explorations of Captain Cook and others in the Southern Ocean. Between 1770 and 1830 the edge of the Antarctic ice appears to have been perhaps a degree of latitude south of its position in the years 1900–50 (Lamb, 1967). On the other hand, the climatic deterioration seems to have persisted rather longer in Antarctica, and to have lasted until 1900 or later (Lamb, 1969). The data for the southern hemisphere in general are still meagre, but Salinger (1976) has been able to show that New Zealand experienced a climatic deterioration starting at about A.D. 1300, which was most severe between 1600 and 1800. This ties in well with the situation in Europe and North America.

One possible major result of the Little Ice Age in lower latitudes was the extensive deposition of alluvium in the Mediterranean valleys. The major phase of aggradation of Levailloiso–Mousterian age during the Würm was followed by a phase of erosion, culminating in the Neolithic. This was ended by the medieval alluviation, which has itself been succeeded by a renewed phase of erosion (Fig. 4.8). Although it has sometimes been maintained that agricultural and pastoral practices may have been responsible for some of these fluctuations, Vita-Finzi (1969) has written that 'A climatic interpretation of historical aggradation in the Mediterranean is favoured by its incidence from France to the Hoggar and from Palestine to Morocco, by the knowledge that medieval Europe experienced a minor Ice Age and other prolonged periods of anomalous climate, and by the inadequacy of other explanations.'

The French economic historian and historical geographer, Braudel (1972) has proposed that the marked climatic decline of the late sixteenth century had various significant effects on life in the Mediterranean lands, and refers to a high incidence of floods on rivers like the Rhone. The Guadalquivir iced over at Seville, and at Marseilles the sea froze in 1595 and 1638. Of more especial economic significance was the series of frosts which killed the olive trees of Languedoc in 1565, 1569, 1571, 1573, 1587, 1595, 1615, and 1642. Likewise, Ladurie (1972) has shown how the fluctuations of the little optimum and of the Little Ice Age affected traditional societies through their influence on both wine quality and on harvest dates. 'In these societies,' he writes (p. 23), 'mainly agricultural, and dominated by the frequently diffi-cult problem of subsistence, the relation between the history of climate and the history of man had, in the short term, an urgency it has now lost.'

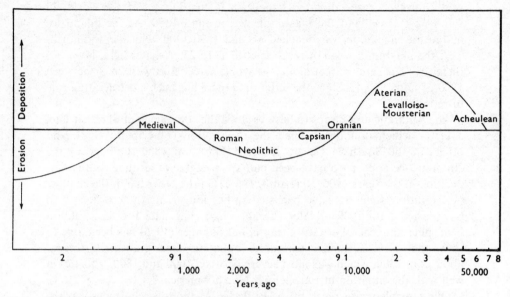

FIG. 4.8. Phases of erosion and alluviation in the Mediterranean valleys over the last 50 000
years (after Vita-Finzi, 1969, p. 92).

The Greenland settlements

Against the background of climatic changes involved in the deteriorating
environment of the medieval Ice Age or neo-glaciation, it is perhaps not
surprising that settlements in highly marginal areas like northern Norway,
Greenland, Iceland, and highland Britain, suffered reverses to their economies, which led to grave social effects, and population decrease.

There are, of course, dangers in any simple deterministic interpretation,
but the climatic effects outlined above should not be ignored. As Utterström
(1955), a proponent of the role of climatic change as a determinant
of economic advance and decline has written: 'Ever since Malthus
and Ricardo, all discussions of the pressure of food supplies have started
from the assumption that population is the active factor and Nature
the fixed. This interpretation, however, can hardly be reconciled with
modern scientific thought, especially if the problem is viewed in the long
term.'

Of particular interest is the question of the settlements in Greenland,
which were established by the Vikings, reached a reasonable size, and then
declined almost catastrophically in the late-medieval period. It has been
argued that both sea conditions and the actual environment in Greenland
itself must have been favourable for the initial settlement to take place.
Drift ice rarely appeared near Iceland and Greenland south of 70 °N. in the
900s and was apparently unknown between 1020 and 1194 (Lamb, 1969).
Other lines of evidence in addition to this one have led Lamb to propose

optimum. Viking burial grounds of the period in question in south-west Greenland, in which graves were dug deep and trees rooted, are now permanently frozen, and this has suggested to Lamb (1966) that since that time, mean annual temperatures may have fallen by 2 to 4 °C. However, after 1410 there was no regular communication between Europe and any part of Greenland, and no one seems to have reached anywhere on the east coast of Greenland from the sea between 1476 and 1822. Gradually the settlements failed, and it seems highly likely that deteriorating climatic conditions would have led to a highly unfavourable environment, and to a growth of sea-ice which would limit both communication with the outside world and restrict fishing activity.

However, Sauer (1968, p. 157) has disputed the role of climatic change in these events and has made his position abundantly clear: 'The arguments that the settlements failed because the climate changed seem incompetent or irrelevant. They failed as far and small outposts that slowly lost the ability to live in the European manner.' He proposed that various other factors could be responsible for the decline, including a deterioration in drainage, a degeneration of the people (the more vigorous and venturesome being lost by emigration), a state of cultural rigidity imposed by the population being too small, too remote, too little diversified in skills, and too unenterprising, a series of raids by European 'pirates', and a lack of wood for building their own ships. The last of these would lead to increased isolation so that life became very unattractive. Nevertheless, certain of these factors, including emigration, isolation, the absence of wood, and lack of enterprise, could be caused by environmental changes. Moreover, the evidence of changes in sea-ice limits cannot be disputed.

Medieval highland cultivation

In much of northern Europe, including Britain, the little climatic optimum and some of the preceding centuries, were times of extension of settlement to highland areas. About the 1230s medieval villages and their strip cultivation systems were spreading so far on to the higher ground as to cause anxiety for the preservation of enough pasture. In Central Norway the limits of settlement, forest clearance, and farming were pushed 100 to 200 m farther up the hillsides and valleys in Viking times (A.D. 800–1000), whilst in many parts of Britain there is evidence of medieval tillage on high ground far above anything that would be reasonable now, even in war time (Lamb, 1966). For instance, the thirteenth-century limit of tillage in Northumberland appears to have been around 300–350 m above sea-level, 120–150 m above the limit of any worthwhile possibility at the present day. This was also a period when medieval vineyards spread into a number of localities in southern and eastern England. After a while, however, these extended settlements underwent a fairly considerable decline, and much of this decline started before the Black Death. Of nearly fifty deserted villages in Oxfordshire and thirty-four in Northamptonshire only about 10 per cent were attributable to the Black Death of 1348. All appear to have suffered serious decline in the years of disastrous summers and famines between 1314 and 1325 (Lamb, 1967).

In Scotland, Parry (1975) has proposed that secular deterioration of climate since the early Middle Ages caused much of the high-lying cultivation in south-east Scotland to become profoundly sub-marginal in the seventeenth century. The consecutive harvest failures of the 1690s and 1780s may have been the immediate stimulus to abandonment, but the response to these stimuli, he maintains, would neither have been so widespread nor so permanent if the potential for cropping in the upland areas had not been so severely reduced over the preceding three centuries. On the basis of the data presented by Lamb and others for the degree of climatic change, and by a study of the present-day climatic limitations on oat ripening, Parry (1975) suggests that in the early medieval warm period the chances of crop failure were only about 1 year in 20 in the Lammermuir area. By the mid-fifteenth century this was down to one year in three, and in all about 4950 ha of land seem to have been abandoned in the Lammermuir–Stow Uplands.

Figure 4.9 plots the progress of the climatic deterioration in that area, using two parameters. One measure of the intensity of summer warmth is the accumulated temperature calculated over a base of 4·4 °C. The isopleth of 1150 day-degrees C showed a marked correspondence with the 1860

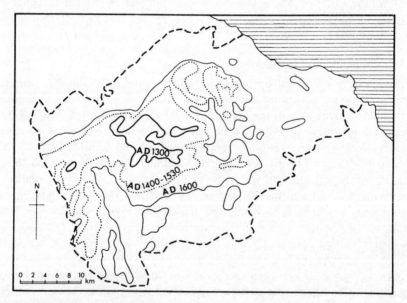

FIG. 4.9. Climatic deterioration 1300–1600, in the Lammermuir-Stow uplands of Scotland, represented by the fall of combined isopleths of 1050 day-degrees C and 60 mm PWS (from Parry, 1975, Fig. 5).

cultivation limit. PWS (Potential Water Surplus) is a measure of summer wetness, expressed as the excess of a middle and late summer surplus (up to 31 August) over an early summer deficit.

In Denmark, the villages most often abandoned were those terminating in '-thorp'. These were just those villages that had been established relatively late—from the tenth to the twelfth centuries (Steensberg, 1951). In Iceland, corn growing decreased greatly not long after 1300, ceased altogether in the sixteenth century, and was not to be re-established again until after the Little Ice Age. It appears to have reached its maximum in the tenth century. In the first half of the fourteenth century the centre of gravity of economic affairs moved away from the interior of the island to the coast, where fishing was to become the main economic activity in place of cereal growing and *vadmal* (homespun) production (Utterström, 1955).

In Sweden, the so-called 'Golden Age' of Gustav Ist ended in the mid-sixteenth century, and this prosperous era was followed by an era from which there is a plethora of reports of natural catastrophes, crop failures, and famine, persisting for a hundred or so years from the 1590s.

Another economic consequence of this phase of deteriorating climate was that the silver mines of Central Europe were subjected to increased water problems after 1300, leading to the closure or rundown of many mines.

Reading for Chapter Four

The Holocene is treated in some detail in both Butzer, K. W. (1972) *Environment and archeology: an ecological approach to prehistory* and in West, R. G. *Pleistocene geology and biology* (1972). A series of papers at the Royal Meteorological Society, *Climate from 8000 B.C. to 0 B.C.* (1966), have been published and contain much useful material. The climatically and man-induced vegetational changes of the British Holocene are exceptionally well treated in W. F. Pennington (1969), *The history of British vegetation*.

The climatic optimum, or hypsithermal, has been discussed by E. S. Deevey and R. F. Flint (1957), 'Post-glacial hypsithermal interval', *Science* 125, 182–4; and the recent work on neo-glaciation which has caused some modification in ideas on the optimum is summarized by G. H. Denton and S. C. Porter (1970), 'Neo-glaciation', *Scientific American* 222 (6), 101–10. Other valuable papers giving more local detail include J. H. Mercer (1967), 'Glacier resurgence at the Atlantic/Sub-Boreal transition', *Quarterly Journal Royal Meteorological Society* 93, 528–33; R. R. Curry (1969), 'Holocene climatic and glacial history of the Central Sierra Nevada', *Geological Society of America Special Paper* 123, 1–47; C. J. Heusser (1961), 'Some comparisons between climatic changes in north-western North America and Patagonia', *Annals New York Academy of Sciences* 95, 642–57; G. D. McKenzie and R. P. Goldthwait (1971), 'Glacial history of the last eleven thousand years in Adams Inlet, south-eastern Alaska', *Bulletin Geological Society of America* 82, 1767–82.

Other papers of general interest which deal with the zonation of the Holocene and its problems include U. Hafsten (1970), 'A sub-division of the late Pleistocene period on a synchronous basis, intended for global and universal use', *Palaeo* 7, 279–96; J. H. Mercer (1969), 'The Allerød oscillation: a European climatic anomaly?', *Arctic and Alpine Research* 1, 227–34; G. Manley (1971), 'Interpreting the meteorology of the Late and Post-Glacial', *Palaeo* 10, 163–75; and, on the British situation in particular, G. Manley (1964), 'The evolution of the climatic environment' in W. Watson and J. B. Sissons (eds.), *The British Isles: A systematic Geography*.

The question of the impact of the Late Holocene climatic changes on man and his settlements has yet to be fully explored, though the work of E. Huntington is still immensely stimulating, notably his *Pulse of Asia* (1907) (Constable, London), and certain sections of *Climatic changes* (1922) (Yale University Press, New Haven), written with S. S. Visher. A useful discussion of Huntington's views in the light of more recent research by Soviet and other workers is J. Chappell's (1970) paper, 'The Pulse of Asia Reconsidered', *Geographical Review* 50, 347–73. The possible impact of the Early Holocene environmental changes has been discussed by C. O. Sauer (1948), 'Environment and culture during the last deglaciation', *Proceedings American Philosophical Society*, 92, 65–77, but the same author (1968) disputes the role of climatic change in the decline of the Greenland settlements in his *Northern mists*.

A contrary view on the role of the medieval fluctuations, especially the Little Optimum and the Little Ice Age in Europe comes from a number of historians and geographers including: A. Steensberg (1951), 'Archaeological dating of the climatic change in north Europe about A.D. 1300', *Nature* 168, 672–4; G. Utterström (1955), 'Climatic fluctuations and population problems in early modern history', *Scandinavian History Review* 3, 1–47; H. H. Lamb (1965), 'The early medieval warm epoch and its sequel', *Palaeo* 1, 13–37, and (1966) 'Britain's climate in the past', *The changing climate*. An excellent French work, now translated, which deals comprehensively with the question of the Little Ice Age, is E. L. Le Roy Ladurie's (1972), *Times of Feast, Times of Famine: a history of climate since the year 1000*.

The massive Pleistocene extinctions of fauna are dealt with in P. S. Martin and H. E. Wright's (eds.) (1967), *Pleistocene extinctions*.

Other reviews of the same theme have been made by J. J. Alford (1971), 'A geographic appraisal of Pleistocene overkill in North America', *Proceedings Association of American Geographers* 3, 10–14, and G. S. Krantz (1970), 'Human activities and megafaunal extinctions', *American Scientist* 58, 164–70.

The role of climatic changes in affecting the prehistory of western America have, in spite of the reservations of geographers with regard to environmentalism, been the subject of a massive recent literature, especially in *American Antiquity*. The following are among the more valuable reviews: C. Irwin-Williams and C. V. Haynes (1970), 'Climatic change and early population dynamics in the south-western United States', *Quaternary Research* 1, 59–71; H. E. Malde (1964), 'Environment and man in arid America', *Science* 145, 123–9; R. B. Woodbury (1961), 'Climatic changes and prehistoric agriculture in the south-western United States', *Annals New York Academy of Science* 95, 705–9; J. B. Griffin (1967), 'Climatic change in American prehistory' in R. W. Fairbridge (ed.), *The encyclopaedia of Atmospheric Sciences and Astrogeology* 169–71; F. Oldfield and J. Schoenwetter (1964), 'Late Quaternary environments and early man on the southern High Plains', *Antiquity* 38, 226–9; D. J. Lehmer (1970), 'Climate and culture history in the middle Missouri valley', in W. Dort and J. K. Jones (eds.), *Pleistocene and recent environments of the central Great Plains* 117–29; and S. C. Jett (1964), 'Pueblo Indian migrations: an evaluation of the possible physical and cultural determinants', *American Antiquity* 29 (3), 281–300. However, much excellent early work was undertaken by Kirk Bryan and is still worth reading. See, for example, his 'Pre-columbian agriculture in the south-west as conditioned by periods of alluviation', *Annals Association of American Geographers* 31, 219–42 (1941).

With regard to the old world there is no comparable body of modern literature, and for a fairly lengthy discussion one has to go back to C. E. P. Brooks (1949), *Climate through the Ages* (2nd edn.), and some of the many papers by Huntington (for a bibliography of his works see S. S. Visher's obituary of him in *Annals Association of American Geographers* (1948), 38, 39–50). However, the following more modern treatments are apposite: H. H. Lamb (1968), 'The climatic background to the birth of civilisation', *Advancement of Science* 25, 103–20; E. Demougeot (1965), 'Variations climatiques et invasions', *Revue Historique* 228, 1–22; and, on the classical world, H. E. Wright (1968), 'Climatic change in Mycenaean Greece', *Antiquity* 42, 123–7; R. Carpenter (1966) *Discontinuity in Greek Civilisation* (Cambridge U.P.). There is also some material on the question of the role which climatic change played in the momentous cultural developments of post-glacial times in the Near East. A moderate climatic interpretation is that of R. Solecki (1963), 'Prehistory in Shanidar Valley, northern Iraq', *Science* 139, 179–83, whilst pertinent data are referred to by H. E. Wright (1968), *Climatic change in the eastern Mediterranean region* (University of Minnesota, Contract Nonr 710 (33), Task no. 389–129, Final report). More general discussions about the relationship of domestication to environmental change include C. Vita-Finzi (1969), in P. J. Ucko and G. W. Dimbleby (eds.), *The domestication and exploitation of plants and animals*, Aldine, pp. 31–4; and E. Isaac, *Geography of Domestication* (1970).

5 Environmental Changes During the Period of Meteorological Records

> Whatever the future may bring, we are justified in say-
> ing that of the endless series of climatic fluctuations that
> have occurred from the beginning of the Earth and that
> will continue in the future, the present one is the first
> that we can measure, investigate, and possibly explain.
>
> H. W. AHLMANN (1953, p. 31).

Changing temperatures in the twentieth century

Although various scholars, especially Brooks, Ladurie, Manley, and Lamb, have adroitly interpreted past climatic conditions from documentary records, thereby greatly adding to our knowledge of climatic conditions in pre-industrial Britain and Europe, it was not until the nineteenth century that there was an organized growth of instrumental observations from stations all over the world.

It is on the basis of these relatively reliable instrumental records that most of our knowledge of the latest chapter of environmental evolution is founded.

Such records, while infinitely more reliable than methods utilized for previous centuries, are not without their limitations; instruments need to be replaced and recalibrated from time to time, and sites and locations are liable to change because of such factors as urbanization or vegetational change. However, by careful selection of the more reliable stations avail-able, and by taking averages for several stations within an area, a valuable picture of changes can be obtained.

The extent of changes in climate over the last 100 years or so is greater than was formerly believed; both temperature and rainfall have shown trends which have led periodically to great fluctuations in glaciers, lakes, and river discharges. A comparison of these climatic changes with others in Britain since the 1690s is made in Figure 5.1. It is, however, dangerous to generalize too much about the nature of the changes on a world basis, as even over quite short distances trends may have been in opposite directions, or may have shown a time lag. Nevertheless, the following are some features that are particularly notable.

First, there was a general phase of warming during a part of the present century, though the degree of change has varied according to latitudinal position, with a tendency for the greatest increases to be in high-latitudes in

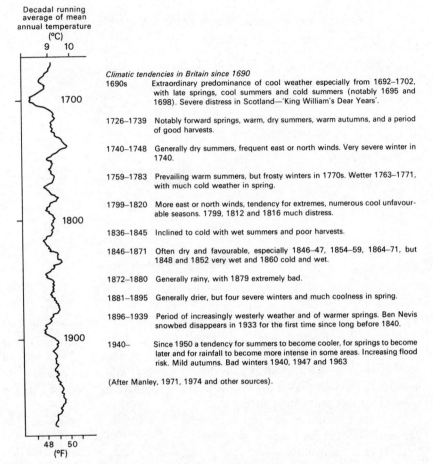

Decadal running
average of mean
annual temperature
(°C)
9   10

Climatic tendencies in Britain since 1690

1690s   Extraordinary predominance of cool weather especially from 1692–1702, with late springs, cool summers and cold summers (notably 1695 and 1698). Severe distress in Scotland—'King William's Dear Years'.

1726–1739   Notably forward springs, warm, dry summers, warm autumns, and a period of good harvests.

1740–1748   Generally dry summers, frequent east or north winds. Very severe winter in 1740.

1759–1783   Prevailing warm summers, but frosty winters in 1770s. Wetter 1763–1771, with much cold weather in spring.

1799–1820   More east or north winds, tendency for extremes, numerous cool unfavourable seasons. 1799, 1812 and 1816 much distress.

1836–1845   Inclined to cold with wet summers and poor harvests.

1846–1871   Often dry and favourable, especially 1846–47, 1854–59, 1864–71, but 1848 and 1852 very wet and 1860 cold and wet.

1872–1880   Generally rainy, with 1879 extremely bad.

1881–1895   Generally drier, but four severe winters and much coolness in spring.

1896–1939   Period of increasingly westerly weather and of warmer springs. Ben Nevis snowbed disappears in 1933 for the first time since long before 1840.

1940–   Since 1950 a tendency for summers to become cooler, for springs to become later and for rainfall to become more intense in some areas. Increasing flood risk. Mild autumns. Bad winters 1940, 1947 and 1963

(After Manley, 1971, 1974 and other sources).

48   50
(°F)

FIG. 5.1. Climatic tendencies in Britain since 1690.

the northern hemisphere. This is illustrated in Table 5.1, where two slightly different interpretations of the degree of change are given. Both agree, however, on the relatively great amount of change north of 60 °N.

Even within such a small country as Sweden the south seems to have warmed up less than the north. The amount of increase in the mean December–March temperature for the years 1901–30 compared with 1859–1900 was 1·2 °C for Haparanda and Ostersund, 0·87 °C for Uppsala, and 0·5 °C for Lund (Liljequist, 1943). But in some situations the latitudinal variation in degree of change suggested above seems not to hold. In the Middle East, for instance the biggest increases in temperature were in the south and not in the north.

The rise in temperature from the 1900s to the 1930s amounted to 0·5 °C at Nicosia, 0·75 °C at Beirut, 0·8 °C at Jerusalem and 0·9–1·0 °C at

Table 5.1
Twentieth-century temperature changes

Warming trend in various latitudes
(i) *After Callendar* (1961)

Lat. range	Mean temperature 1921–50 minus Mean temperature 1891–1920 (°C)
73 –60 °N.	+0·83
60 –25 °N.	+0·39
25 °N.–25 °S.	+0·17
25 °S.–50 °S.	+0·14

(ii) *Schell* (1961)

Lat. range	Mean temperature 1921–1940 minus Mean temperature 1901–1920 (°C)
60 –80 °N.	+1·43
40 –60	+0·43
20 –40	+0·31
0 –20	+0·27
0 –20 °S.	+0·24
20 –40	+0·24
40 –60	+0·10
60° –	–0·40

Alexandria, Cairo, and Khartoum. Equally, a study by Wexler (1961) of conditions at Little America, Antarctica suggests that between 1911 and 1958 there was a general trend upwards of 2·6 °C between 1912 and 1957, a finding which seems to conflict with that of Schell (1961), who, as shown in Table 5.1a thought that in the far south of the southern hemisphere changes were either minimal or negative.

Urbanization almost certainly accounts for some local variations. The Japanese evidence, for example, suggests that between 1910 and 1950 the most rapid rises of temperature occurred in large cities such as Tokyo, Osaka, and Kyoto with amounts of 0·9, 0·6, and 0·9 °C respectively. Japanese scientists found that rural stations showed a rise but that it was considerably smaller, and suggested that 60 per cent of the increased temperature in the great cities could be ascribed to increased urban influence on the micro-climate rather than to any general change in climatic conditions (Fukui, 1970).

Although most areas, both in the northern, and in the southern hemispheres, showed a general rise in mean annual temperatures in the first half of this century, there is evidence to suggest that some seasons may have been relatively warmer, and some relatively cooler. This was, for example, the case in East Asia, where mean January temperatures in Hongkong fell 0·8 °C between 1884–1910 and between 1911–1940, while mean July temperatures rose 0·2 °C. In Kyoto, over roughly the same period, the January fall was 0·2 °C and the July rise 0·9 °C. Almost the reverse has been the case in Central Europe as Table 5.2 shows. This again indicates the danger of excessive generalization.

Table 5.2

*Changes in mean temperatures (°C),
1881–1910 compared with 1911–40*

	January	July
Vienna	+1·74	−0·50
Zurich	+1·72	−0·34
Utrecht	+1·01	+0·10
Uppsala	+1·30	+1·22
(Kraus, 1956)		

The East Asian patterns probably resulted from the effects of an increased winter pressure over the continent which led to more frequent and vigorous cold northerly winds over the China Coast and Japan. In Europe, more cyclonic conditions led to warmer winters and cooler summers. In other words, a less 'continental' regime seems to have been established, and there was a greater frequency of westerly weather types.

The date when the amelioration in temperature reached its maximum has varied from area to area. In the British Isles 1931–40 was the warmest decade in the extreme north-west (Stornoway), whereas the warmest decade did not occur until 1943–52 in the south-east (Kew). The maximum was reached in the 1930s in the Middle East, but in the 1920s in Alexandria (Rosenan, 1963). In Japan the warming went on to 1961, but declined thereafter. Table 5.3 indicates the dates of the warmest and coldest decades, together with their temperature characteristics for Europe.

One consequence of the warming trend can be seen when one looks at the

Table 5.3

Changes in temperature conditions (April to June) at European stations since 1860

Station	Warmest decade	Mean temperature (°C)	Coldest decade	Mean temperature (°C)	Difference (°C)
Angmagssalik	1926–35	2·14	1899–1908	0·09	2·05
Vestmannø	1889–98	7·69	1948–58	5·63	2·06
Spitzbergen	1951–60	−2·99	1912–21	−5·99	3·00
Haparanda	1945–54	6·22	1873–82	4·08	2·14
Bodø	1945–54	6·53	1873–82	5·29	1·29
Helsinki	1945–54	9·22	1873–82	7·13	2·09
C. England	1943–52	11·82	1879–88	10·46	1·36
De Bilt	1940–9	14·12	1951–60	11·71	2·41
Zurich	1942–51	13·85	1879–88	12·34	1·51
Milan	1943–52	18·98	1879–88	16·80	2·18
Barnaul	1938–47	11·36	1882–91	8·61	2·75

(After Harris, 1964)

Table 5.4

Snow, ice, and frost frequencies in the nineteenth and twentieth centuries

(a) *Frost days, ice days, and cold days in Sweden*

No. of	1861–70	1871–80	1881–90	1891–1900	1901–10	1911–20	1921–30	1931–40
Frost days [1]	121·8	122·3	123·4	124·7	125·0	115·4	117·2	103·4
Ice days [2]	55·9	58·2	57·2	57·9	56·3	55·9	57·2	47·1
Cold days [3]	43	48	24	33	8	19	21	19

1 = days with a minimum temperature $< 0\,°C$
2 = days with a maximum temperature $< 0\,°C$
3 = days with a maximum temperature $< 10\,°C$
(After Liljequist, 1943)

(b) *Snow frequencies at selected stations*

Location	1895–6 to 1937–8 (1)	1895–6 to 1937–8 (2)	1938–9 to 1961–2 (1)	1938–9 to 1961–2 (2)
Lerwick	10·3	23	24·6	83
Tynemouth	7·4	16	10·9	33
Cambridge	7·7	14	12·4	33
Ross on Wye	5·4	9	9·7	21
Freiburg (SW. Germany)	23·2	70	31·8	92

(1) = Average number of days with snow lying
(2) = Percentage of winters with 15 days or more with snow lying
(After Lamb, 1969)

(c) *Days of ice cover in Norway and Sweden*

Lake	1900–	1910–	1920–	1930–	1940–
Femund (Norway)	176·3	158·8	168·2	156·4	161·8
Mjosa (Norway)	—	71·6	65·2	22·8	49·8
Rössvatn (Norway)	—	164·4	159·0	138·9	144·6
Bolmen (Sweden)	—	105·2	89·4	86·0	93·9
Siljan (Sweden)	131·1	120·2	110·0	108·5	102·0
Storsjön (Sweden)	168·4	155·8	149·9	145·5	144·6
Mean value	—	129·33	123·62	109·68	116·1

(From data in *World Weather Records* processed by author)

(d) *Dates of the first and last snowfall in London from 1811 to 1960*

	Autumn	Spring
1811–40	18 Nov.	22 Apr.
1841–70	21 Nov.	17 Apr.
1871–1900	23 Nov.	12 Apr.
1901–30	25 Nov.	15 Apr.
1931–60	8 Dec.	1 Apr.

(After Manley, 1964)

(e) *Characteristics of April weather at Newark, Nottinghamshire*

	1946–66	1967	1968	1969	1967–9
Average daily minimum (°C)	4·9	3·8	3·2	2·2	3·07
Monthly minimum (°C)	0·2	−5·2	−6·5	−3·4	−5·03
Number of frosts	1·1	4	9	9	7·33
Days with sleet or snow	1·3	1	3	3	2·33

(After Lyall, 1970)

dates of the first and last snow falls in London from 1811 to 1960. As Table 5.4(d) shows, whereas in the early years of the nineteenth century the mean dates of the first and last falls were separated by over 150 days, by the period 1931–60 this figure had declined to only 113 days. Even when one compares 1931–60 to 1901–30, the period during which one might expect snow was reduced by around four weeks. This may partly result from the effects of urbanization.

Another consequence of the greater warmth was that the length of ice cover on rivers and lakes in high-latitudes declined appreciably until the 1930s or later. Some data for Norway and Sweden are shown in Table 5.4(c), and of the Lakes considered, Mjosa in Norway is the one which shows the greatest decline in ice cover, with an average of 71·6 days of ice per year for the decade after 1910 falling to only 22·8 days in the 1930s.

In recent years there have been signs in some areas that conditions have become somewhat colder since the optimal years for temperature which lasted in many areas during the twenties and thirties.

This is illustrated for the Northern hemisphere up to 1970 in Fig. 5.2F. where pentad means are plotted as deviations from the 1880–4 pentad. Most of the United States had cooler mean temperatures in 1961–70 than in 1931–60 (Kalnicky, 1974). Almost all the eastern U.S.A. averaged at least 0·5 °C cooler.

At Oxford, from 1910 to 1939 there were only nine winters colder than the average and only three of these were more than 1 °C colder. Between 1939 and 1965, on the other hand, 16 winters were colder than the average and 5 of them were particularly cold. Similarly from 1926 to 1938 no year exceeded the mean in terms of snow cover: between 1939 and 1965, 14 did (Smith, 1967). Also in the late 1960s there was a distinct trend towards colder weather during the spring season, a critical time for horticulture and agriculture, as is illustrated by the data from Newark, Notts. (Lyall, 1970). Daily minimum temperatures, monthly minima, and the number of frosts have changed markedly (Table 5.4e) in the month of April.

Similar data are available for Oxford where, on a temperature basis associated with the first occurrence of five days with temperatures high enough for plant growth, an operational definition of spring has been made (Fig. 5.2E). For the first fifty years of the period 1869–1970 the ten-year moving mean date of the first day of spring (as defined) fell between 13 and 22 March, but as a result of the warming trend of the 1930s and 1940s the date around 1940 was as early as 3 March. However, since 1961–70 the date of spring has become progressively later, the growing season has started later, and the mean date for the start of spring has been around 19 March. All the ten-year means for 1922–60 were on or before 13 March, except for one, whereas during the last five years of the period 1869–1970 the ten-year means have all been after 13 March (Davis, 1972).

Similarly dramatic changes have also taken place in the growing season in

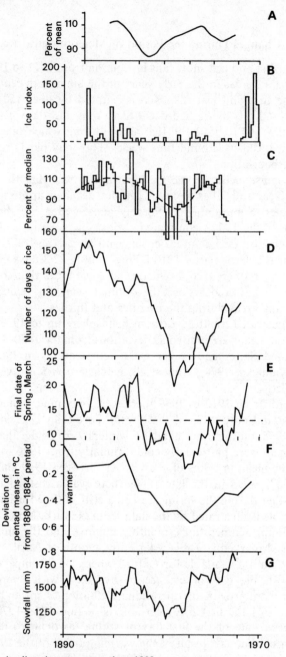

Fig. 5.2. Changes in climatic parameters since 1900.

A. 20-year running means of the mean winter–spring rainfall at 14 stations in North Africa and the Middle East (from Winstanley, 1973).

B. Extent of ice off Iceland (duration in weeks multiplied by the number of areas with ice along the coasts) (from Schell, 1974).

C. Variation in annual runoff in the United States as a whole (from Leopold *et al.*, 1964, p. 62). The dotted line represents a generalized trend.

D. 10-year running means centred at date given of number of days for season with ice in the Baltic at Stugsund (from Davis, 1972).

E. Final date of spring at Oxford, England, represented by a 10-year running mean (from Davis, 1972).

F. Temperature changes of the northern hemisphere, shown by pentad means expressed as deviations from the 1880–4 pentad (from Kalnicky, 1974).

G. 10-year moving mean of snowfall (in mm) at the Blue Hill Observatory, Massachusetts, U.S.A. (from data in Conover, 1967).

the United States corn belt (Brown, 1976). This period (defined as the number of days between the last killing frost in the spring and the first in the autumn) has declined by between 22 and 43 days at stations in the corn belt over the last 30 years, using 11 year running averages. As the average growing season in the corn belt is now 165 days, little further decrease can be tolerated as the minimum requirements for hybrid corn are 100–30 days.

The Baltic Sea's ice cover has shown a broadly comparable trend to that of springs at Oxford (Fig. 5.2D) with a sharp decline between 1895 and the mid 1930s being followed by a steady climb in the number of days of ice cover to the mid-1960s.

Equally, water temperatures over most of the North Atlantic north of 40° have shown some decline in recent years, with sharp falls of up to 2·5 °C in the 1950s in the west Atlantic between 40° and 60 °N. The effects if such a trend continues will probably be detrimental to the north European fishing industry (see p. 157).

Of no less importance for transport undertakings and the like, is the increased incidence of snow which has become apparent since the last temperature optimum. The figures in Table 5.4b give an indication of the way in which, since the late 1930s, the number of days with snow and the frequency of long spells of snow have increased: at Lerwick in the Shetland Islands the number of days of snow more than doubled.

Equally, in New England, snowfall records from 1885–6 up to 1965–6 indicate that snowfalls declined considerably until around 1940, with low quantities of snowfall for most of the 1920s and 1930s. However, since 1940 (see Fig. 5.2G) there has been a marked increase in snowfall frequencies and quantities, and the decadal moving means of the late 1950s were the highest since records began. The ten-year moving mean centred on the winter of 1931 was 117 cm of snowfall; by 1959 there was a ten-year moving mean snowfall value of 183 cm.

In the eastern Canadian Arctic (Baffin Island) changes in both temperature and precipitation have been sufficient to lead to an increase in snow cover and the development of permanent snowbanks and 'glacierets' during the 1960s (Bradley & Miller, 1972). A marked decrease of temperature by 2 °C in the summer melting season, and higher winter precipitation, have resulted in snowbanks encroaching on 25 mm diameter lichens, lichens that to reach that size must have been snow free for about the previous forty years.

In conclusion it is possible to say that when one compares temperature conditions of 1900–19 with those of 1920–39 one finds that about 85 per cent of the earth's surface experienced warming trends in mean annual temperature, whereas when one looks at temperature data for the period between 1940 and 1960 about 80 per cent of the total earth surface has probably been involved in a net annual cooling (Mitchell, 1963). Only a few areas, such as the western United States, New Zealand, south-east Canada, eastern Europe, the Pacific coast of Asia, the Brazilian plateau, and various portions of the western Indian Ocean have continued to show a net warming since 1940, and it is still not yet clear how widely this situation still persists into the 1960s and 1970s. In New Zealand, however, the warming still seems to be going on (Salinger and Gunn, 1975) (Figure 5.3), as it does over much of Australia (Tucker, 1975).

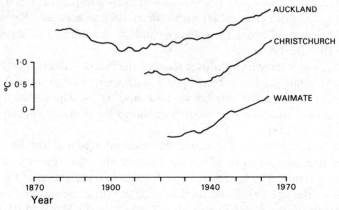

FIG. 5.3. 20-year running mean of mean annual temperature for typical New Zealand stations (after Salinger and Gunn, 1975).

Changes in rainfall

The changes in rainfall that have taken place during the period of instrumental records are as difficult to generalize about as are temperatures. However, the changes have been considerable. Over much of the tropics and subtropics for instance precipitation dropped to a very low level in the 1920s and 1930s after being at a maximum during the 1890s and 1900s (Table 5.5a, b, c). The drop in precipitation in the Middle East from the peak, to the 1920s and 1930s, accounted for 12–18 per cent of the mean at Nicosia, Beirut, and Tel Aviv, 30–44 per cent at Haifa, Jerusalem, and Alexandria, and no less than 77 per cent at Cairo. The absolute amounts of decrease in rainfall were 2 cm of mean annual rainfall at Cairo, 6–11 cm at Alexandria, Tel Aviv, Beirut, and Nicosia, and 19–20 cm at Jerusalem and Haifa (Rosenan, 1963). The data for the eastern United States show a similar fairly marked decrease in the first part of the twentieth century compared to the last forty years of the nineteenth century (Table 5.5d).

In the Mediterranean area, Gibraltar showed a decrease in rainfall of about 11 per cent (100 mm) between 1881–1910 (943 mm) and 1911–40 (842). Further east, Aden, in line with other stations in the Middle East showed a 34 per cent decrease (19 mm) over the same period. After 1940 the rainfall at most places increased again, but there was a dry spell around 1960.

Table 5.5
Rainfall fluctuations in the tropics, U.S.A. and Australia

(a) *Tropical Rainfall*

	1874–98	Percentage deviations from 1881 normal of mean annual rainfall	1907–11
Barbados	+14		− 9
Bogota	+10		− 3
Colombo	+ 4		− 8
Freetown (1875–99)	+11		−12
Georgetown (Queensland)	+20		− 8
Havana	+10		− 5
Honolulu	+13		−12
Recife	+42		−11
Townsville	+17		− 4
Trinidad	+10		− 7
Vizagpatan	+10		0

(b) *Table average rainfall as per cent of 1881–1940*

	1861–80	1881–1900	1901–20	1921–40	1941–60
E. U.S.A. (7 stations) (30–43 °N)	125	109	91	99	108
E. Australia (4 stations) (19–38 °N)	113	111	96	93	108

(c) *Mean rainfall in Queensland* (cm/year)

	Georgetown	Townsville	Gilbert River
1872–96	95·25	137·2	97·3
1911–40	73·2	90·7	68·3

(d) *U.S.A. rainfall changes* (percentage deviations from the mean)

	Mean 1881–1940 (cm)	1861–1900	1901–40
Charleston	112·3	+21·4	−7·8
Washington	103·9	+ 5·8	+1·0
New York	108·7	+ 4·3	−2·9
Albany	83·1	+17·1	−2·9
Boston	100·8	+12·3	−2·3

(After Kraus, 1954, 1955 (a), 1955 (b))

Some of the most detailed studies of rainfall changes have been made in Britain and these show up both the quantitative variability in trends between different regions and the temporal variability in maxima and minima (Gregory, 1956).

Those parts of Britain exposed to westerly influences showed a sequence with falling totals from 1881 until 1892–1901; increases until 1909–18; stability until 1922–31 at this high level; and falling totals until 1950. In general most British stations showed a rise in rainfall until the early 1920s, and then a fall, the fall generally starting in 1923–32, but there were regional differences in the beginning of the rise, the rate and mode of the rise, and the date at which most maximum values were recorded. However, taking the period 1900–59 in northern England, a contrast arises between those areas of rapid orographic uplift of prevailing westerly air-masses and juxtaposed areas on lee sides. The former shared a significantly large increase in rainfall amounting to 15 per cent in the Manchester lowland and the Lake District and 10 per cent over Rossendale, the Bowland Fells, and the head of the Lune Valley. Actual decreases occurred on the lee of the Pennines in the Eden Valley and in the Slaithwaite area. Equally Slaithwaite had its maximum from 1910 to 1919 while the far west near the coast had a maximum from 1923 to 1932 (Barrett, 1966).

Rainfall changes of the nineteenth and twentieth centuries in the low latitudes

As already mentioned, in many parts of the tropics and subtropics the period corresponding to the warming phase of higher latitudes was a time of decreased precipitation. This is, for instance, shown by the data for eastern Australia presented in Table 5.5(b) and (c). In all about 2·5 m km² of Australia showed significantly decreased precipitation for the period 1911–40 compared with 1881–1910. Only 0·25 m km² showed an increase. The decreases were especially severe in the semi-arid area near Bourke (Fig. 5.4) where a decrease of 75 mm in mean annual rainfall took place at this time. This was equivalent to a regression of some 100 km in the isohyets (Gentilli, 1971).

In the dry zone of south Asia, another region where any deterioration in rainfall would have severe human implications, there was a noticeable change in rainfall around 1890 to 1895. Conditions had been relatively wet in the 1880s and 1890s, but then there followed a period of low precipitation, with precipitation in the driest decadal period being generally only between 52 and 69 per cent of that for the wettest decade of this century. This change in regime is well illustrated by the graphs of the ten-year moving means of precipitation at Lahore and Karachi (Pakistan), Jaipur (Rajasthan), and Agra (United Provinces) (Fig. 5.5). After about 1940 or 1945 there seems to have been a return to more positive rainfall conditions.

This is mirrored in the record for central and southern Africa, where after relatively moist conditions in the pre-Boer War period, an abrupt change in

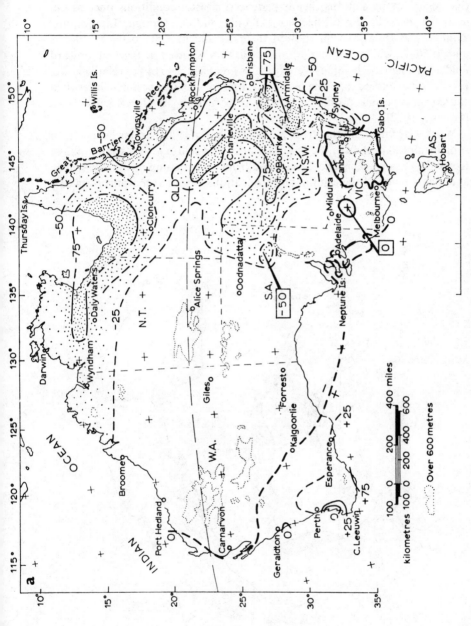

Fig. 5.4. Shifts of climate in Australia since 1881. (a) Isopleths of the yearly rainfall difference (1881–1910 minus 1911–40)

rainfall conditions came in the mid 1890s which lasted until the 1930s. In that region many stations then experienced moister conditions once again, and, for instance, Port Elizabeth, Luanda, Ndola, Mongu, Livingstone, Zomba, and Bulawayo all showed their decadal maxima between 1947–8 and 1956–7, whilst Kimberley and Salisbury showed a marked upward tendency. These high values were often as great as those of the relatively wet pre-Boer war period, and the high discharges of the Zambesi which resulted caused unexpected problems in the construction of the Kariba Dam (Goudie, 1972).

Work elsewhere in the tropics, as Table 5.5(a) illustrates, suggests a similar rainfall decline in the first quarter of this century for a wide range of tropical locations. This decline resulted from a shortening of the wet season and a narrowing of the rainfall belts. Locations on desert margins showed a

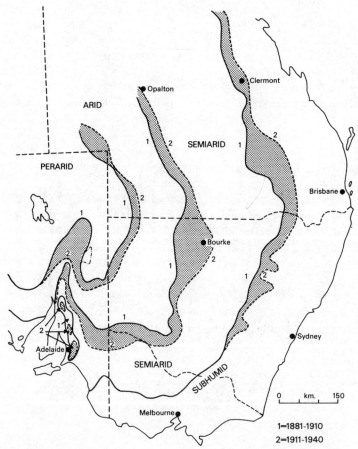

Fig. 5.4. (b) Shifts in the climatic belt boundaries. 1 is 1881–1910 and 2 is 1911–40 (after Gentilli, 1971).

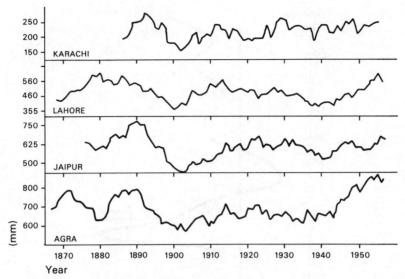

FIG. 5.5. 10-year moving means of precipitation at Indian dry zone and Pakistan desert stations (from Goudie, 1972).

relatively greater decline. In West Africa, for example, Bathurst in the Gambia showed a more substantial change than Freetown in Sierra Leone.

The data from Table 5.5(b) from the eastern parts of the U.S.A. and Australia further illustrate the decline in totals since the 1880s to a minimum in the period 1900–40. They also indicate that some rise has taken place since 1941. Indeed, north-eastern Australia has recently experienced some remarkable floods in normally arid districts.

During the 1960s there appears to have been a considerable change in precipitation conditions over wide areas, a change that may equal in importance that of the late 1890s. More years of records will be required before this can be stated categorically, but the available evidence is weighty. The precipitation changes have been of both a positive and of a negative character.

A particularly sharp increase in rainfall took place in the equatorial parts of East Africa. The rainfall figures for the thirty-six-month period up to mid-1964 were, according to Lamb (1966), 130 or 140 per cent of the 1931–60 averages, and in places were over 250 per cent. A broadly similar pattern of high precipitation levels in the equatorial zone occurred in 1970–2 (Fig. 5.6).

This had very serious consequences in terms of changed discharges for the Nile and higher levels (q.v.) for the East African lakes. The Nile, for instance, at its outlet from Lake Victoria, had a mean discharge for the sixty-three-year period before 1962 of around 600 m³/sec. Since 1963 this has more than doubled to a figure of around 1200–1300 m³/sec.

However, this marked rainfall increase did not occur universally in the

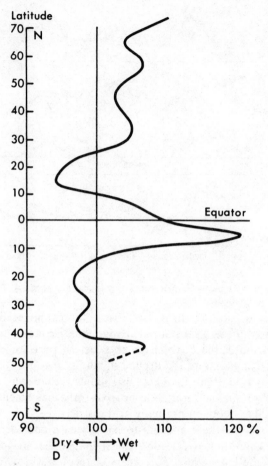

Fig. 5.6. Graph of annual precipitation 1970–2 as percentage of 1931–60. The Sahelian
drought north of the Equator has a parallel at 20°–30 °S. Between the two dry zones
is a relatively wetter equatorial zone (after Lamb, 1974, in Rapp, 1974).

tropics, and the zones between 10–15° and 30 °N. and S. of the Equator
(Fig. 5.6) were drier than usual, with catastrophic drought conditions being
experienced in, for example, Botswana, where, during parts of the mid-
1960s, the bulk of the population were forced to subsist on famine relief.
Similarly, the Dead Sea, unlike the equatorial lakes, had a very high level
between 1898–1932 of − 393·1 to − 395·3 m and then dropped very rapidly
between 1957 and 1963 to as little as − 398·8 m (Klein, 1965) as a result of
both increased irrigation in the Jordan valley and low rainfall.

A recent study of these rainfall trends over North Africa, the Middle
East, and north-west India shows a very similar picture (Table 5.6). From
Mauritania across to north-west India the summer monsoon rainfall has
decreased steadily by more than 50 per cent since 1957. It is this phen-

Table 5.6

Five-year running mean percentage of normal summer monsoonal seasonal rainfall centred on 1957 and 1970.

Location	1957	1970
Bikaner, India	114	71
Jodhpur, India	115	68
Khartoum, Sudan	122	80
Agades, Niger	130	44
Tessalit, Mali	140	63
Gao, Mali	114	75
Nouakchott, Mauritania	106	74
Atar, Mauritania	121	52
Mean	120	66

(After Winstanley, 1973)

omenon, superimposed on the increasing population pressures of many developing countries, which has led to the severe and much-publicized spectre of drought in West Africa and in India. However, although certain climatologists (see, for example, Winstanley, 1973) have postulated that this drought is part of a longer term progression towards aridity in monsoonal areas, there is as yet very little statistical evidence to support the contention (Bunting *et al*, 1976). Indeed, the West African drought of the late 1960s and early 1970s was not unparalleled (Rapp, 1974). The rainfall for the period 1907–15 seems to have been as weak, and the discharge of the Senegal and Upper Niger in the period 1910–14 was inferior to that of 1968–72. Neither on the northern or the southern side of the Sahara has analysis of available climatic data shown an upward or downward trend that is statistically significant.

Another area where rainfall totals have declined markedly over the last decade or so is north-central Chile. This decline, identified by Lloyd (1973), has intensified an already downward trend which started in the mid-1940s. That part of the economic structure of the area which is dependent upon water has, as a consequence, been seriously affected. River discharges have declined, and this is reflected in the agricultural output of the area. There has been a gradual decline in the production of barley and maize since 1954 while the staple wheat crop was maintained until 1964 since which time a steep decline has set in. The 1968–9 wheat crop was only 17·5 per cent of the 1961–2 crop. Cereals have suffered particularly in that there is a low financial return on water used for cereal cultivation, and so scarcer water resources have been utilized preferentially for the cultivation of higher value crops such as grapes.

The important groundwater systems in this area of Chile are alluvial and confined to the valleys. They are small units, and these too have been seriously depleted, so that yields from wells have dropped. The copper processing concerns in this area depend on water, and they, like agriculture, have been hit by groundwater contraction. Thus, for example, the completion of a new copper plant in Combarbala has been delayed, in that adequate supplies are no longer available nearby. In Domeyko records show that the wells supplying the copper-processing plant yielded 17 l/sec. in 1942. These wells now only yield 2 l/sec. and production in the plant has dropped to 40 per cent of capacity.

Changing tropical lake levels

One of the most interesting examples of environmental change in the present century has been the fluctuating level of lakes in the tropics. In particular many Equatorial lakes in Africa showed dramatic increase in level in the early 1960s, which led to the flooding of port installations, deltaic farming land, and the like (Butzer, 1971). This rise contrasted sharply with the frequently low levels encountered in the previous decades.

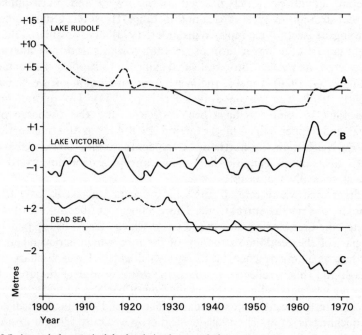

Fig. 5.7. Annual changes in lake levels in the twentieth century.
(A) Lake Rudolf, East Africa, showing the low levels from 1930 to 1960, and the relatively high levels around 1900 and since 1960.
(B) Lake Victoria, East Africa, showing the marked 'stepwise' rise in level since 1960.
(C) Dead Sea, showing the generally positive hydrological budget from the start of the century to around 1930, and the highly negative budget during the 1960s, resulting in part from the diversion of Jordan waters for irrigation.
(From Butzer, 1971, Figs. 5–5 and 5–4.)

Lake Malawi (Nyasa) seems to have reached a minimum around 1927–9, Lake Tanganyika was very low in the 1920s and again between 1948 and 1956, and Lake Chilwa was 9 m lower in level than it had been at the time of David Livingstone. Lake Victoria's lowest level was reached in 1922, while Lake Naivasha showed a very sharp progressive fall after 1938. The relatively dry phase which seems to have been the dominant reason for these low levels in the 20s and following decades started to develop in the 1880s in the basin of lakes Nyasa, Tanganyika, and Victoria but rather later (around 1898) in the Rudolf and Stefanie basins. The lakes studied by the colonial scientists were thus very different from those described by the great explorers and may have contributed towards the concept of progressive desiccation and desertification which concerned foresters and others in Africa between the wars.

In the 1960s, the level of the lakes of East and Central Africa rose sharply. Lake Tanganyika stood about 3 m higher in 1964 than it had in 1960, Lake Victoria rose by 1·5–2·2 m (Fig. 5.7), and rises of 2·3 m have been recorded by lakes Baringo, Nakuru, and Manyara. Lake Rudolf began to rise 4 m in late 1961 and submerged over 300 km^2 of the Omo Delta. In 1970 the author noted that trees in the Galla lakes basin south of Addis Ababa were submerged by several feet of water. Further south by 1963 Lake Nyasa had risen by almost 6 m compared with its minimum height of 1915.

River discharge fluctuations

The publication by UNESCO in 1971 of many carefully selected discharge records for some of the world's major rivers enables one to see the way in which river discharges have fluctuated in response to the changes in both temperatures and precipitation. The rivers selected by UNESCO are ones for which there are long, reliable records and where the direct effects of human interference (such as irrigation, diversion of drainage, and the like) are not too significant. The data have been analysed by the author to obtain ten-year moving means of the mean monthly annual discharges in m^3/s (Goudie 1972). Thirty rivers from the northern hemisphere were selected for this study on the basis of the length and continuity of their records.

The graphs of the variability in ten-year moving means, some of which are reproduced in Fig. 5.8, show that considerable fluctuations have taken place, and a better impression of this can be gained by examining the ratios of the maximum to minimum decadal discharges for the period of observation. The mean ratio for the thirty rivers was 1·78, though there was a range from 1·19 to 6·49. This is equivalent to having had minimum mean ten-year periods with discharges only a little over 50 per cent of the maximum mean ten-year periods.

Analyses of the dates of the maximum and minimum mean ten-year periods do not suggest any general progressive decline in discharges as some

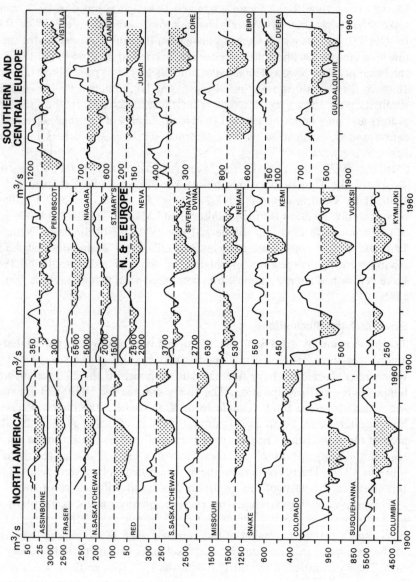

Fig. 5.8. Data on river discharge (10-year moving means of the mean annual monthly discharge) for selected stations (after Goudie, 1972). Periods of low discharge are shaded.

of the proponents of the concept of progressive desiccation might have hoped. However, of the thirty rivers considered, no fewer than seventeen showed their minima between 1935–6 and 1945–6 (mid-years of the ten-year periods). The maxima are much less clustered, though nine rivers showed maxima between 1948–9 and 1958–9, and many others showed something of an upward trend in this period after low levels in the 1930s and 1940s.

It will be noticed how the North American rivers, of which thirteen are used in Fig. 5.8 showed in almost all cases a general decrease during the first three or four decades of this century, with the lowest discharges being attained during the so-called 'dust bowl' years of the 1930s. This was a time of higher than average temperatures and lower than average precipitation over much of North America (see also Fig. 5.2C).

The U.S.S.R. rivers, Neman, Neva, and Severnaya-Dvina also show their minimum flows in the early 1940s and in this way they compare closely with the Finnish rivers, Kymijoki, Vuoksi, and Kemi. However, some of the rivers from further south or west in Europe (Labe, Danube, Duera, Guadalquivir, Jucar, and Ebro) show either their minima or a discharge trough in the late 1940s or early 1950s. The Russian and Finnish discharges also exhibit a secondary trough between 1910 and 1920.

In this respect they are comparable to the two rivers considered from West Africa, the Niger and the Senegal. They show low discharges in the period 1910 to 1920, and again, after higher discharges, low discharges in the early 1940s. These periods of low flow were comparable to those of the late-1960s and early-1970s (Fig. 5.9).

The discharge of the White Nile has also fluctuated markedly. The minimum ten-year period of discharge as determined at Lake Albert occurred in 1926–7, a time when the lake levels in East and Central Africa were also low. For the decade centred on 1926–7 the discharge was only 19·2 milliards

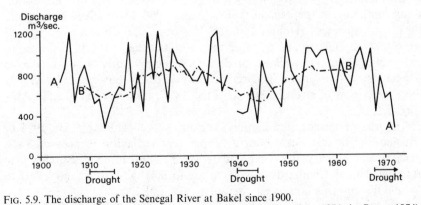

FIG. 5.9. The discharge of the Senegal River at Bakel since 1900.
    (a) The record of mean annual discharge (after Sircoulon, 1974, in Rapp, 1974). Three drought periods of 1910–14, 1940–4, and 1968–72 are evident. Data for 1939 are missing.
    (b) 10-year moving mean of discharge (after Goudie, 1972).

of cubic metres per year while for the wet decade centred on 1915–16 it was up to 28·0.

Glacial fluctuations in the twentieth century

The last Little Ice Age, as already noted, had largely ended by the end of the nineteenth century. Since then many mountain glaciers have been in retreat, often at very fast rates. Some have disappeared altogether, and in the tropics, for example, where recent glacial histories seem to be comparable to those of higher latitudes, six glaciers have disappeared from Ruwenzori since they were first described by explorers in the middle of the last century, and at the present rate Mount Speke will be completely deglaciated within less than four decades (Whittow *et al.*, 1963). The Elena Glacier on Ruwenzori averaged a retreat of 5 m/yr. between 1900–52, but this accelerated between 1952 and 1960 to a rate that reached between 6·5 and 25 m per year. In the Himalayas most glaciers are also retreating, with an average rate of 40·2 m per annum since 1885 for the Pindari glacier in the Kumaon region (Ahmad and Saxena, 1963).

Similar rates of recession have occurred in high latitudes. Between 1899–1901 and 1936, for example, the Lady Franklin Glacier of Svalbard retreated about 2·5 km (Thorarinsson, 1940). The Jostedals Glaciers of Norway on average retreated about 160 m between 1910 and 1921. They then advanced 60 m until 1930 and retreated at a steadily increasing rate so that 58 m of recession was recorded by 1946 (Ahlmann, 1948). Equally the Svartisen Ice Cap, was reduced in area from 468·1 km² (1894–1905) to 400 km² (1965) (Theakstone, 1965). The Jostedals advances were paralleled in the Alps where retreat was arrested about 1906, and advance developed, culminating between 1916 and 1920, with almost 75 per cent of Alpine Glaciers participating. However, after 1926 the bulk of the Alpine glaciers were in retreat (Fig. 5.10), largely because of a reduction of cyclonic conditions compared to the decades 1886–95 and 1906–15 which were followed by glacier advances (Hoinkes, 1968). However, the advances of 1915–25 were generally inadequate to bring their snouts back to the 1895 positions.

In spite of several decades of decreased temperatures in Europe most Alpine glaciers continued their retreat into the 1960s but by the late 1960s there were signs of strong advance in the Swiss, Italian, and Austrian Alps (Fig. 5.11).

The loss of glacial area in these regions was considerable. By 1925 or therabouts, for instance, about a 25 per cent reduction in area of Swiss glaciers had taken place compared to the position in 1875. Similarly, in Italy, of the 239 Lombardy glaciers at the turn of the century, no less than 66 had disappeared completely between 1905 and 1953.

The Alaska record shows the same basic features. Since their Little Ice Age maximum, for example, Herbert and Eliot Glaciers had receded 3·2 km and 610 m respectively. The Muir ice front has been particularly con-

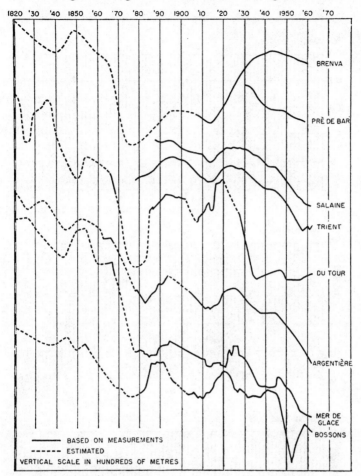

1820 '30   '40 1850 '60 '70 '80 '90 1900 '10   '20   '30   '40 1950 '60 '70

BRENVA

PRÈ DE BAR

SALAINE

TRIENT

DU TOUR

ARGENTIÈRE

MER DE GLACE

BOSSONS

——— BASED ON MEASUREMENTS
----- ESTIMATED
VERTICAL SCALE IN HUNDREDS OF METRES

FIG. 5.10. Changing frontal positions of larger glaciers of the Mont Blanc Massif, the Alps (from Grove, J. M., 1966).

spicuous by its retreat. Its area has been reduced by 450 km² between the 1890s and 1940s; during the period 1899–1913 alone it receded about 12·9 km, and the Muir Glacier of the 1890s had been dismembered into twelve separate glaciers. The Lemon Creek Glacier recession since 1759 has been well documented (Heusser and Marcus, 1964), and it is interesting to see that the rate of retreat, which had been very rapid since 1891, was somewhat reduced in the period 1902–29, which seems to correspond partially with the period of partial advance in the European Alps (Table 5.7a).

Not all glaciers, however, have followed the general pattern outlined above. In Alaska, for example, the Taku Glacier, in the Juneau ice field, advanced 5·6 km in forty-eight-years while the other glaciers were retreating (Lawrence, 1950). Its pattern appears in Table 5.7b. Equally the Crillon

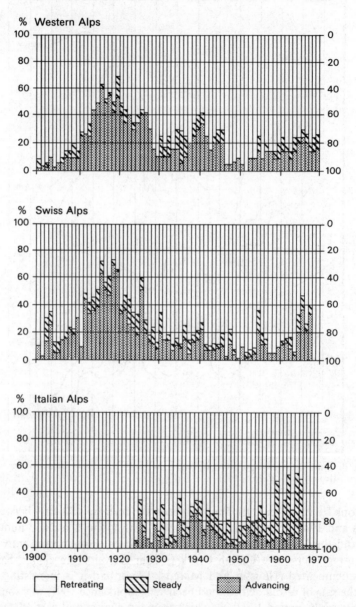

FIG. 5.11. Variations in glacier snouts 1900–70 for the Alps (from Vivian, 1975, Fig. 32).

Table 5.7

Fluctuations in Alaskan Glaciers

Glacial fluctuations since the Little Ice Age

(a) *The history of Lemon Creek Glacier (Alaska), 1759–1958*

Date	Per cent area loss	Cumulative per cent area loss	Recession rate m/yr.
1759	0·8	0·8	—
1759–69	0·7	1·5	12·5
1769–1819	0·9	2·4	3·5
1819–91	0·9	3·3	3·8
1891–1902	4·7	8·0	61·4
1902–19	2·3	10·3	4·4
1919–29	1·4	11·7	7·5
1929–48	9·7	21·4	32·9
1948–58	4·1	25·5	37·5

(After Heusser and Marcus, 1964)

(b) *The advances of Taku Glacier, Alaska, 1900–52* (m)

1900–19	117
1909–29	83
1929–31	183
1931–7	159
1937–41	160
1941–8	83
1948–50	51
1950–2	91

(After Field, 1954)

Glacier in south-east Alaska advanced at a rate of 28 m per year between 1894 and 1933, and showed an over-all advance of 4·5 km since 1786 (Goldthwait *et al.*, 1963). In the European Alps the Brenva Glacier has shown an advance between 1925 and 1940 when all its neighbours were in retreat (Fig. 5.10).

The reasons for such anomalous movements are many. In the case of the Brenva it was due to the effects of a spread of protective avalanche debris which descended on to the Glacier in 1920 from Mont Blanc de Courmayeur (Grove, 1966). The Taku and Crillon Glaciers of Alaska owe their anomalous advance to the fact that they are nourished from a higher source area. Taku's source begins at 1800 m, for example, whereas nearly all the other ice streams begin at 1200–1500 m (Heusser *et al.*, 1954). The advance of the Jan Mayen Glaciers since the early 1950s is due to a greatly increased cyclonic precipitation in these high latitudes which has more than outweighed any increased temperatures associated with the cyclonic activity (Lamb *et al.*, 1962). It has been estimated that precipitation in the 1950s has almost doubled in comparison with the 1920s. Indeed, conditions which cause one glacier to advance may cause another to retreat. In western Norway, for example, advection of warm moist cyclonic air generally plays the greater part in ablation, whereas in Sweden, radiation, which is often reduced during cyclonic conditions, appears to be even more important.

Ground laid bare by the retreat of ice caps and glaciers under the pressure of the relatively warm conditions of the first half of the twentieth century, become colonized by vegetation in stages. In Alaska three main stages have been found in the succession (Lawrence, 1958). The first, or *pioneer* stage sees initial colonization by hardy *Rhacomitrium* mosses, a gradual increase in perennial herbs, particularly the broad-leaved willow herb and horsetail, and finally the establishment of *Dryas drummondii*, the latter being a low-growing evergreen under shrub.

The second stage, the *thicket* stage, witnesses the appearance of dwarf, creeping willows which together with the *Dryas* lead to an increasing degree of shade which leads to the gradual suppression of the shade-intolerant mosses and herbs of the *pioneer* stage. Towards the end of this stage there is a dominance of shrubs of willow (*Salix spp.*) and alder (*Alnus spp.*). Alder is important because it is an important source of soil nitrogen enrichment, and thereby provides improved soil conditions for the next stage in the succession.

The third stage in the succession is the *forest* stage when there is a dominance of Sitka spruce (*Picea sitchensis*), and later a mixture of spruce and hemlock (*Tsuga spp.*).

This sequence whilst only strictly applicable to the Glacier Bay area of Alaska nevertheless gives a broad view of the situation that probably existed in most areas during deglaciation.

There is some evidence that as a result of the halt to the warming trend in parts of the northern hemisphere, the period of glacial retreat is either over or showing signs of coming to an end (Meier, 1965). In the Western U.S.A., for example, the picture in the Olympic Mountains and Cascade Range of Washington State was that more than fifty glaciers were showing enlargement as early as 1953–5 (Lawrence and Lawrence, 1961). There are also signs that 'active' glaciers in Spitzbergen, like the Hans Glacier, are beginning to advance (Kosiba, 1963).

Some consequences of current climatic changes for ocean conditions

The consequences of current climatic change are not fully brought out simply on the basis of a study of average temperatures and precipitation totals. One of the most important indicators of, and consequences of, environmental change was the change in ice cover which took place in the high latitudes of the northern hemisphere.

For example, the warming led to a general diminution in the ice cover in the Arctic Seas which had major implications for navigation. Off Iceland, both in the 1860s and 1880s, there had been on average between twelve and thirteen weeks of ice in a year. By the 1920s the incidence was down to only 1·5 weeks per year, though by the decade 1947–56, because of the temperature fall already described, this had increased slightly to 3·7 weeks in the year. Equally, the area of drift ice in the Russian sector of the Arctic was reduced by no less than 1 million km² between 1924 and 1944 (Diamond, 1958). The ice also tended to become less thick, so that whereas Nansen found that the average thickness of ice in the Polar Sea was 365 cm in the period 1893–6, the *Sedov* expedition of 1937–40 found that it was down to 218 cm (Ahlmann, 1948). Iceberg frequencies off Newfoundland also declined. The annual average for 1900–30 was 432 whereas for 1931–61 it was 351, a decrease of 19 per cent (Schell, 1962). The coast of Greenland also became less subject to ice as illustrated by the frequency of years in which the Polar Ice, which comes round Cape Farewell, reached as far north as Godthaab. From 1870 to 1879 it was over 70 per cent; since 1910 it has always been less than 25 per cent (Beverton & Lee, 1965).

As a result of the improved ice conditions, the shipping season for the coalfields of West Spitzbergen lengthened from three months at the beginning of the century to about seven months in the 1940s.

The changes of sea-temperature associated with these changes in ice cover were of a high order. The changes were generally positive, though some areas, notably those affected by the Irminger current off Iceland, cooled (Brown, 1953). Off the Kola peninsula the water temperatures in the early twenties were 1·9 °C higher than they had been twenty years earlier. Similarly, between 1912 and 1931 the sea-water temperatures off north western Spitzbergen rose by 1·5 °C. The sea around Iceland has shown a fairly continuous upward temperature trend in most, but not all sectors, amounting to about 1·5 °C between 1925 and 1960. In general the majority of areas showed their greatest rise after 1916–20.

After 1960, however, in line with the observed temperature diminutions and other indications of environmental change, the amount of sea-ice off Iceland showed a substantial increase, attaining levels that had been unknown for over forty years (Fig. 5.2b). Indeed the values for 1970 were the highest of the century, though there has been some decline since then.

Faunal changes in the northern seas

The effects of these temperature increases on the fishing industry are now well documented. The colonization of the west Greenland continental shelf by the Cod (*Gadus morhua*) from Iceland is the best example of the response to the general warming trend. Before 1917, except probably for short periods during the nineteenth century, only small local fiord populations of cod occurred in Greenland. After 1917 large numbers of adult fish appeared off the south-west coast as far north as Freiderickshaab (72 °N.) and they migrated north through 9° of latitude in twenty-seven years (Ahlmann, 1948). As a result 10 000 tons of cod were landed in Greenland in 1948 compared with only 5 tons in 1913. The haddock (*Melanogrammus aeglefinus*), and the halibut (*Hippoglossus vulgaris*), showed a similar northward movement towards both Greenland and to Novaya Zemlya. Between 1924 and 1949 swordfish, pollack, twaite shads, and dragonets were recorded for the first time off Iceland. Amongst the species that appeared more frequently were mackerel, tunny, horse mackerel, conger, basking shark, thorn-back ray, mullet, forkbeard, saury pike, and rudderfish. The great silver smelt and the Greenland shark extended their range (Cushing, 1976). On the other hand, there was a striking response of typically cold water forms such as the white whale (*Delphinapterus leucas*) and the capelin (*Mallotus villosus*). Their southerly limits have contracted ((Beverton & Lee, 1965).

The Baltic Sea also benefited from the climatic amelioration. Its salinity increased as a result of increased frequency of south-easterly winds which tended to increase the outflow of brackish surface-water from the Baltic and brought about a corresponding increase in the compensating inflow of saline North Sea water along the sea-bottom. Over the period 1933–9 salinities were up to 1·7 per cent higher than during the period 1923–32. High salinity levels improve spawning conditions for cod, and this has led to an enormous, perhaps twentyfold, increase in the abundance of the Baltic cod, which now supports a major fishery (Beverton & Lee, 1965). A rise in salinity of 0·1 per cent also occurred in the NE. Atlantic (1919–38 compared with 1902–17) (Weyl, 1968).

The results of the amelioration may also include the dramatic decline of the Plymouth herring fisheries of the English Channel, the herring fisheries of the Firth of Forth and the haddock fisheries of the North Sea. The Channel fisheries have been partially replaced, after 1935, by warmer water forms, especially the pilchard (*Sardina pilchardus*) and the cuttlefish (*Sepia officinalis*). A decrease in the amount of zooplankton and of nutrient salts in sea-water, especially in the winter months, was recorded in the Plymouth area during the warming period. In general, however, the results of the increased temperatures were beneficial for the north European fishing industry. A reversal to conditions comparable to those at the beginning of the century is now taking place, and the west Greenland cod fishery has now nearly disappeared. Already cod, ling, and haddock have returned to the Plymouth area of south-western England, and the boreal barnacle (*Balanus balanoides*) has become greatly prominent along the shoreline. Many changes in fish numbers must also be put down to overfishing, but the role of climate has been an important one (Russell *et al.*, 1971).

Faunal and floral changes in the northern hemisphere

Land flora and fauna of Northern Europe have also showed changes in their distribution, though the biological consequences of changes in sea-temperature are likely to be more definite than the corresponding changes on land. The sea is more uniform so that it might be expected that temperature and salinity would offer the main restrictions to the spread of oceanic species. The influence of man also tends to be less direct, though of course, as just noted, overfishing has had near catastrophic effects on the distribution of certain coastal fisheries.

In Finland the polecat (*Mustela putorius*) began spreading into the country about 1810 and by the late 1930s had occupied the whole south Finnish interior to about 63 °N. (Kalela, 1952). The colder the winter and the greater its snowfall the more difficult it is for this mammal to find its natural food of small rodents, frogs, and the like. In NE. Greenland the Musk Ox had plenty of food, from 1910 onwards, and its numbers increased markedly (Vibe, 1967). Likewise, the Roe Deer, common in south and central Scandinavia in the years before the Little Ice Age was almost extinct there by the early nineteenth century, only to reappear and spread strongly northwards after 1870. Also in Finland some permanently resident birds such as the partridge, to which severe snow is inimical, the tawny owl, and many species of tit, have extended northwards (Crisp, 1959).

In Iceland and Greenland the distributions of birds give an equally clear indication of the effects of the amelioration (Harris, 1964). For example, the fieldfare (*Turdus pilanis*) was unknown in Greenland and Jan Mayen before 1937, but breeding is now established there. Starlings (*Sturnus vulgaris*) arrived in Iceland in 1935 and became permanent in 1941. Swallows of one species (*Hirunda rustica*) appeared in the Faeroes and Iceland in the 1930s. The white-fronted goose and the long-billed marsh wren have commenced to breed in Greenland, and some species like the mallard and long-tailed duck, which were summer visitors, now remain throughout the year. However, the reduction in numbers of the little auk provides a clear example of how an improving climate may have a directly adverse effect on a species. It feeds on small Crustacea such as mysids, which are particularly plentiful in the surface water near the sea-ice front. As the sea-ice front retreated northwards from Iceland the birds in the Icelandic colonies had to fly over larger distances for food, and the colonies were thereby gradually deserted (Crisp, 1959).

These environmental changes have also not been without their economic implications. The rise in temperature has increased the growing season of crops. In Finland, for instance, Helsinki showed, for the period 1934–8, twenty-three more days per year without frost compared to the mean for 1901–30. Over the same period there were also twenty-two more growing days (days when the average temperature persists above 5 °C) (Keranen, 1952). The data for Sweden show a similar trend (Table 5.4a). Trees grew at a greater rate in Arctic Finland, and the Scandinavian countries experienced an extension of rye, barley, and oat growing which was not occasioned solely by the breeding of more tolerant strains.

In other parts of the world the bio-geographic and economic consequences of climatic change are perhaps less well-established, though the dust bowl of the 1930s in the United States was not a result solely of bad agricultural practice. From 1931 to 1940 the temperatures in the western Great Plains were over 0·75 °C above the average while precipitation was over 15 per cent lower. One authority has written that in the Great Plains 'the great drought of 1934–40 was so severe and so prolonged that only fragments of the major consociations of which true prairie is composed remained at its conclusion. Moreover, invading xerophytic grassland dominants had taken over most of the area and with other species had almost overwhelmed any climax survivors' (Whyte, 1963).

Another type of invasion took place in the Cascade Range of Washington and Oregon where subalpine meadows were subjected to a massive invasion of a variety of tree species, especially *Abies lasiocarpa* and *Tsuga merten-siana*. The invasion was especially intense during the very warm years 1928–37, but with the cooling since that time there has been relatively very little expansion of trees into the meadowland. This striking environmental

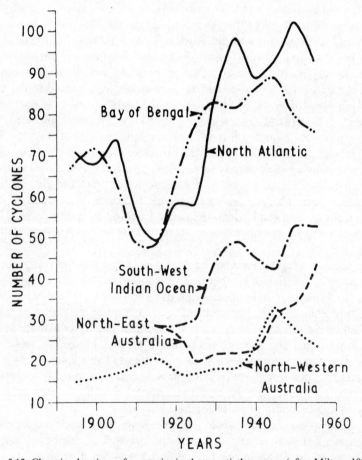

Fig. 5.12. Changing hurricane frequencies in the twentieth century (after Milton, 1974).

change took place over a wide region, precluding the operation of local factors such as reduction in burning in accounting for the phenomenon, and it is highly probable that it is the increase in the length of the snow-free period which has been the critical factor affecting tree establishment in the subalpine meadow environment (Franklin *et al.*, 1971).

Also in the American context it is interesting to note that hurricane frequencies in the American tropics have increased considerably so that while there were fifty in the period 1911–20 there were over 100 in the period 1950–60 (see Dunn and Miller, 1960). In view of the damage inflicted by hurricanes such an increase is probably economically and socially highly significant (see Fig. 5.12). Also, since the start of this century, the location of hurricane tracks has undergone some change which seems to be correlated with changes in sea-water temperatures (Riehl, 1956). In the early years of the century most recurvatures in the hurricane tracks took place to the east of Florida. They then shifted westwards to the Gulf between 1910 and 1920 (a period of relatively cool sea-water temperatures); later (after 1920) they returned at first to Florida and adjoining waters, and finally (in the 30s and 40s) to the west Atlantic. In all, the shift in the average longitude of hurricane track recurvature near latitude 25 °N was no less than 20°. In general, when sea-water temperatures decreased the hurricane tracks migrated westward, and when temperatures increased they returned eastward.

The increase in hurricane frequencies in the Americas appears to have been followed in the Indian Ocean, off Australia (Milton, 1974) (Fig. 5.12), and in Japan (Fujita, 1973).

The dual role of climatic change and human interference

Frequently, however, it is difficult to isolate the various factors that have led to noticeable environmental changes; a specially serious problem is to try and identify whether it is natural environmental change or whether it is man which has led to some particular developments.

Within England a good example of this is provided by the increased flood risk experienced in the River Severn at Shrewsbury, and the Wye at Hereford over recent decades (Howe *et al.*, 1966).

During the period 1911–40 a flood height of 5·1 m was to be expected only once in twenty-five years; during the period 1940–64 this height was reached once every four years. To put it another way, the flood height with a twenty-five-year return interval in 1911–40 was 5·1 m, while in 1940–64 the flood height with a similar recurrence level had reached 5·9 m. The Wye in Herefordshire (Fig. 5.13a) showed a similar trend. Researches have demonstrated that both man-induced and climatic changes have led to this severe hydrological change. On the one hand man has increased the rate of flood runoff by draining the peat in the Welsh uplands and so increasing the drainage density in the catchment area of the Severn. On the other hand examination of precipitation records at Lake Vyrnwy in mid-Wales

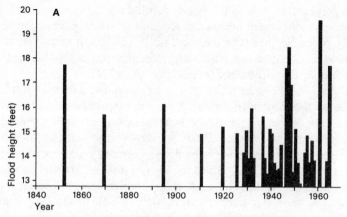

River Wye: recorded flood levels at Wye Bridge, Hereford

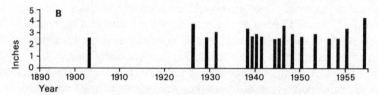

Lake Vyrnwy: frequency of daily rainfalls of at least 2·5 inches

Fig. 5.13. Changes in flood levels and daily rainfall levels for the River Wye and Lake Vyrnwy, western Britain.

indicates a marked increase in the frequency of daily rainfalls greater than 63·5 mm since 1940 (Fig. 5.13b).

It is not yet clear whether such a change in intensity/frequency relationships in recent years is countrywide, but the Welsh data are supported by those from Oxford (Fig. 5.14), where the return interval of large daily falls has shortened appreciably. For the period 1881–1905 the return period for a storm of just over 50 mm was about thirty years, but for the most recent period studied, 1941–65, the return period for the same size of daily fall had dropped to little more than about five years (Rodda, 1969). Should this change be established as being of wide extent it will be extremely significant not only in terms of climate, but in terms of erosion in upland catchments, and from the point of view of flooding and water resources. A greater incidence of floods would have serious economic consequences, not only in terms of damage, but also because the criteria for the design of bridges, spillways, culverts, and similar constructions would need to be revised to deal with the extra risk. Much more work needs to be done to see whether there has been a significant change of more than local dimensions and then to study its implications.

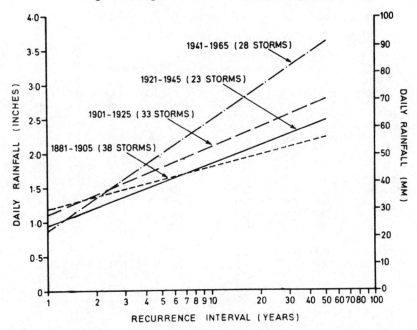

FIG. 5.14. Changes in the magnitude-frequency relationships for daily rainfall amounts at Oxford exceeding 25 mm (from Rodda, 1969, Fig. 8).

In the western United States there has been a considerable and complex debate about the causes of arroyo cutting or gully initiation which peaked in the 1880s (Cooke and Reeves, 1976). Once again a main problem has been whether it was man or climate which was the dominant factor. Overgrazing by cattle and other animals introduced in large quantities by man may well have played a significant role, but Leopold (1951) pointed out that compared with the present day there were more large storms and fewer small ones in parts of the south-west in the 1880s. A few very strong storms would have major consequences for runoff, particularly if storms were less well spaced, for this would reduce the vegetation cover and so would increase the erosion risk. Thus without any great change in mean annual precipitation totals, a change in the frequency and size of storms could have major effects on hydrology, siltation, erosion, and vegetative cover. The figures for three characteristic stations gives a picture of the change in rainfall characteristics that has taken place since 1850. It will be noted that mean annual totals have not changed greatly, but the number of small showers has increased markedly as the frequency of intense storms has decreased (Table 5.8) (Leopold, 1951).

One further demonstration of the complex relationships between changes resulting from the activity of man, and changes resulting from natural causes can be seen in the recent history of the Caspian Sea. Since 1929 this

Table 5.8

Detailed rainfall characteristics of SW. U.S.A.

| Station | Years | (Average number of days of rainfall per year) | | | Mean rainfall total for period (mm) |
		0·25–12·6 (mm)	12·7–25·3 (mm)	25·4+ (mm)	
Santa Fe	1850–80	59·7	5·7	1·96	371
	1881–1910	80·8	5·5	1·07	366
	1911–30	82·8	5·3	1·20	363
	1931–48	71·5	5·1	1·05	378
Las Cruces	1850–80	23·8	3·8	1·65	201
	1881–1910	31·7	4·7	1·17	224
	1911–30	42·1	3·9	0·50	213
	1931–48	42·0	3·3	0·67	218
Albuquerque	1850–80	21·2	1·3	0·40	203
	1881–1910	30·5	2·9	0·71	196
	1911–30	47·0	2·6	0·56	216
	1931–48	58·3	3·4	0·67	229

(After Leopold, 1951)

large inland sea has fallen by three metres in level (Fig. 5.15). This has resulted largely because of a reduction in river inflow from the Volga, which, on average, contributes 80 per cent of the over-all surface discharge to it. Thus over the period from 1929 to 1965 the annual loss of water exceeded the annual gain by around 26 km³, though the deficit was at its greatest between 1930 and 1945 when there was an average annual deficit of 50 km³. The economic consequences of this change in inflow and in level were marked and generally adverse. The Caspian fishing industry diminished as a result of the increasing salinity and the reduction in the numbers of shallows, and fish production declined from around half a million tons per year between 1925 and 1935 to only 82 000 metric tons in the period 1965–8. Moreover, the quality of the fish also declined, with the proportion

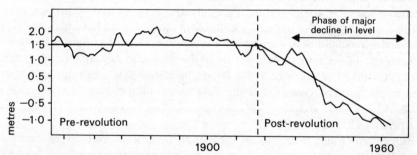

FIG. 5.15. Annual fluctuations of level (in m) of the Caspian Sea, illustrating the rapid post-revolutionary downward trend.

represented by prized types such as sturgeon, white fish, salmon, and herring dwindling particularly strongly (Micklin, 1972).

The causes of this phenomenon are both natural and man-induced. The climatic factor may well have been predominant, particularly in the earlier phases. In the years before 1929 the airflow over the European U.S.S.R. was predominantly westerly but this changed to a chiefly meridional and easterly pattern during the 1930s and 1940s. As a result the number of depressions penetrating from the Atlantic dropped, whereas the frequency of dry anti-cyclones from the Arctic and Siberia went up fairly substantially during the winter season. Lake and river (see p. 151) discharges declined. However, especially over the last twenty-five years, reservoir formation, irrigation, and municipal and industrial withdrawals have been of major importance.

One final illustration may be given of the difficulties that can sometimes arise in a consideration of the different processes that can be involved in some environmental changes. In the Maasai Amboseli Game Reserve in East Africa there has been an extensive loss of the fever tree (*Acacia xan-thophloea*) woodlands over the last two decades, and this has been accom-panied by a shift towards a more arid habitat. There are various alternative hypotheses that could be put forward for explaining this marked change in habitat: it could conceivably result from overgrazing by Maasai livestock; alternatively the death of the trees could be a consequence of elephant destruction resulting from the disruption of their local populations by human encroachment; or, some other factor, such as the marked changes in rainfall regimes that have taken part in East Africa in recent years, could have been responsible.

The hypothesis that overgrazing by Maasai livestock has been respons-ible is at first sight attractive in that such pressures have been recognized in other areas. However, the progressive increase in the number of dead trees towards the centre of the lake basin in the Game Reserve was negatively correlated with the density of livestock. Moreover, the area of highest woodland mortality has been a wildlife sanctuary from which livestock have been excluded since 1961, before the major decline of the trees. Conversely, the basin-edge woodlands which have suffered least loss have, for various reasons, been the areas of highest settlement and cattle density for decades. With regard to elephant damage, this does appear to be conspicuous, but there were also many trees which, though undamaged by elephants, were also dead.

Thus neither of these two hypotheses appears fully to fit the facts. A recent study by Western and Praet (1973), on the other hand, has shown effectively that it is an increasing degree of soil salinity that is the problem. The dead trees are associated with areas of high soil salinity. The most likely explanation for this is an upward shift in groundwater levels of some 3·5 m between 1961 and 1964. This elevation in the water-table caused the capil-lary fringe to introduce a high level of soluble salts into the rooting layer, causing death through physiological drought. The rise in lake levels and of rainfall in the early 1960s is the basic cause of the shift in the groundwater levels (see p. 149).

Conclusion

The great weight of data now available on the question of trends and fluctuations in the twentieth century has led to a great change in attitudes to climatology. As Lamb remarked (1966), 'Not so very long ago . . . climate was widely considered as something static, except on the geological time-scale, and authoritative works on the climates of various regions were written without allusion to the possibility of change, sometimes without mention of the period to which the quoted observations referred.' As Lamb and many others have now shown this static attitude of the 'old climatology' must now be replaced by the dynamic attitude of the 'new climatology'.

Another major conclusion one can draw is that the fluctuations have been of sufficient length and degree to have important consequences, some of which have direct economic significance. This can be emphasized when one looks at agricultural productivity in a marginal area such as Iceland. In the late 1950s hay-yields in Iceland (Bryson, 1974) averaged 4·33 tonnes/ha with the application of 2·83 kg/ha of nitrogen fertilizer. In 1966 and 1967 the yields averaged only 3·22 tonnes/ha with the application of about 70 per cent more fertilizer. The mean temperatures of the warm half of the year of the two periods were 7·65 and 6·83 °C respectively. The climatic reduction of yield overshadowed the expected technological increase.

Reading for Chapter Five

The nature of the climatic fluctuations of the twentieth century have been discussed in a number of useful reviews. The works of H. H. Lamb are notable, particularly 'Climate in the 1960s with special reference to east African lakes', *Geographical Journal* 132 (1966), pp. 183–212, 'Britain's changing climate', ibid. 133 (1967), 445–68, and 'Climatic fluctuations' in H. Flohn (ed.), *World survey of climatology* vol. 2 (1969), pp. 173–249. On a world basis various papers by E. B. Kraus give a good survey; 'Secular changes of the standing circulation', *Quarterly Journal of the Royal Meteorological Society* 82 (1956), 289–300; 'Secular variations of east coast rainfall regimes', ibid. 81 (1955), 430–9; and 'Recent climatic changes', *Nature* 181 (1958), 666–8. Other regional studies include that for South Africa by J. H. Vorster (1957), 'Trends in long range rainfall records in South Africa', *South African Geographical Journal* 39, 61–6, that for Australia by E. L. Deacon 'Climatic change in Australia since 1880', *Australian Journal of Physics* 6 (1953), 209–18 that for New Zealand by M. J. Salinger and J. M. Gunn (1975), 'Recent climatic warming around New Zealand', *Nature* 256, 396–8, that for the Middle East by N. Rosenan (1963), 'Climatic fluctuations in the middle east during the period of instrumental record', *Arid Zone Research* 20, 67–73, that for Japan by E. Fukui (1970), 'The recent rise of temperature in Japan', *Japanese Progress in Climatology*, 46–55, and that for America by R. A. Kalnicky (1974), 'Climatic changes since 1950', *Annals Association of American Geographers* 64, 100–12. Very detailed local studies which illustrate the complexity of spatial variations, include those by E. C. Barrett (1966), 'Regional variations of rainfall trends in northern England, 1900–1959', *Transactions, Institute of British Geographers* 38, 41–58, S. Gregory (1956) 'Regional variations in the trend of annual rainfall over the British Isles', *Geographical Journal* 122, 346–53, and P. B. Wright (1976) 'Recent climatic change' in T. J. Chandler and S. Gregory (eds.), *The climate of the British Isles* (Longman, London). The best single review of the controversial question of whether or not deserts are expanding because of natural climatic changes is by A. Rapp (1974), 'A review of desertization in Africa—water, vegetation, and man', *Secretariat for International Ecology, Stockholm, Report* No. 1.

Some consequences of these climatic changes have been examined in the above references, especially by H. H. Lamb. One of the most important papers is that by H. W. Ahlmann (1948) 'The present climatic fluctuation', *Geographical Journal* 112, 165–95. The biological consequences of the changes in Europe have been examined in numerous works. Particularly useful are D. J. Crisp (1955), 'The influence of climatic changes on animals and plants', ibid. 125, 1–19, C. Vibe (1967), 'Arctic animals in relation to climatic fluctuations', *Meddelelser Groenland* 170, No. 5, 227 pp.; G. Harris (1964), 'Climatic changes since 1960 affecting European Birds', *Weather* 19, 70–9; O. Kalela (1952), 'Changes in the geographic distribution of Finnish Birds and mammals in relation to recent changes in climate', *Fennia* 75, 38–51, F. S. Russell *et al.* (1971), 'Changes in biological conditions in the English Channel off Plymouth during the last half century', *Nature* 234, 468–70, R. J. H. Beverton and A. J. Lee (1965), 'Hydrographic fluctuations in the north Atlantic and some biological consequences', in C. G. Johnson and L. P. Smith (eds.), *The biological significance of climatic changes in Britain*, pp. 79–107; and K. Williamson (1976), 'Recent climatic influences on the status and distribution of some British birds', *Weather* 31, 362–84.

Hydrological changes are imperfectly written up, though long term discharge records for many of the world's major rivers are in UNESCO's *Discharge of selected rivers of the World*, 1971, 194 pp. A useful summary of the East African evidence appears in Butzer, K. W. (1971), *Recent history of an Ethiopian delta*, Chicago. Some discussions of the American evidence appears in L. B. Leopold *et al.* (1964), *Fluvial processes in Geomorphology*, and L. B. Leopold (1951), 'Rainfall frequency: an aspect of climatic variation', *Transactions American Geophysics Union*, 32, 347–57. Another useful treatment, concentrating on the Nile, is that by H. E. Hurst *et al.* (1965), *Long-term storage: an experimental study*, 145 pp.

Conditions in the North Atlantic are discussed in some of the papers and books already cited. P. R. Brown (1953), 'Climatic fluctuation in the Greenland and Norwegian seas', *Quarterly Journal Royal Meteorological Society* 79, 272–81, and I. I. Schell (1962), 'On the Iceberg severity off Newfoundland and its prediction', *Journal of Glaciology* 4, 161–72, contain some useful data.

The literature on glacier fluctuations is massive. Some necessary theoretical considerations are discussed by H. C. Hoinkes (1968), 'Glacier variation and weather', *Journal of Glaciology* 7, 3–20, J. F. Nye (1969), 'The advance and retreat of glaciers', *Weather* 24 (12), 501–12, and (1960), 'The response of glaciers and ice sheets to seasonal and climatic changes', *Proceedings of the Royal Society A*, 256, p. 559, and by J. H. Mercer (1961), 'The response of fiord glaciers to changes in the firn limit', *Journal of Glaciology* 3, 850–64. Only a few regional treatments can be given here: East Africa, J. Whittow *et al.* (1963), 'Observations on the glaciers of the Ruwenzori', ibid. 4, 581–616; Mont Blanc, J. M. Grove (1966), 'The Little Ice Age in the massif of Mont Blanc', *Transactions Institute of British Geographers* 40, 129–43; Norway, W. H. Theakstone (1965), 'Recent changes in the glaciers of Svartissen', *Journal of Glaciology* 5, 411–31; Spitzbergen, A. Kosiba (1963), 'Changes in the Werenskiold Glacier and Hans Glacier in SW. Spitzbergen', *Bulletin International*

Association of Scientific Hydrology 8(1), 24–35; Iceland, S. Thorarinsson (1940), 'Recent glacier shrinkage and eustatic changes of sea-level', *Geografiska Annaler* 22, 131–59; Greenland, A. Weidick (1963), 'Glacier variations in west Greenland in Post-glacial time', *Bulletin International Association of International Hydrology* 8, 75–82; and Alaska, R. P. Goldthwait *et al.* (1963), 'Fluctuations of the Crillon Glacier System, south-east Alaska', ibid. 8, 62–74, and C. J. Heusser and M. G. Marcus (1964), 'Historical variations of Lemon Creek Glacier, Alaska and their relationship to the climatic History', *Journal of Glaciology* 5, 77–86. Data on fluctuations in glaciers appear in P. Kasser (1967), *Fluctuations of glaciers 1959–1965* UNESCO, and P. Kasser (1973) *Fluctuations of glaciers 1965–1970*, UNESCO.

6 Sea-level Changes of the Quaternary

> Flooding of the continental shelves by the sea was certainly the most important geologic event of recent time. It initiated the building of modern deltas, many coral reefs, alluviation of river valleys, and the formation of existing beaches and barrier islands. Doubtless, it also had far-reaching effects on climate and the migrations of marine and terrestrial organisms, including man.
>
> N. D. Newell (1961, p. 87)

The importance of sea-level changes

The climato-vegetational changes of the Quaternary era were only equalled in importance by the relative sea-level changes that took place, though these themselves were partially caused by climatic factors. Other contributing factors included tectonic and orogenic forces, isostatic forces, local compaction of sediments, and loading of sediments into coastal basins of sedimentation. It is also possible to categorize the changes according to whether they were of a world-wide nature and involving sea-level changes (eustatic changes) or of a local nature, and involving changes in land levels (tectonic changes).

The effects of such changes can be seen along most shorelines. Where there are stranded beach deposits, marine shell beds, and platforms backed by steep cliff-like slopes one has evidence of emerged shorelines. One also often has evidence of submerged coastal features such as the drowned mouths of river valleys (rias), submerged dune chains, notches and benches in submarine topography, and remnants of forests or peat layers at or below present sea-level. Many coasts show evidence of both emergent and submergent phases in their history.

Table 6.1 attempts to categorize and list the various causes of sea-level change according to whether they are dominantly worldwide or local. The eustatic types of sea-level change will be discussed first in that they have general significance, while the anomalies on this general pattern caused by local factors, such as isostasy, orogeny, and epeirogeny, will be discussed second.

Eustatic factors

Although glacio-eustasy is the most important of the eustatic factors that have affected world sea-levels during the course of the Quaternary, it is worthwhile to look at some of the other minor eustatic factors which play a

Table 6.1

Factors in sea-level change

Eustatic (World-wide)	Local
Glacio-eustasy	Glacio-isostasy
Infilling of basins	Hydro-isostasy
Orogenic-eustasy	Erosional and depositional isostasy
Decantation	Compaction of sediments (autocompaction)
Transfer from lakes to oceans	Orogeny
Expansion or contraction of water volume	Epeirogeny
because of temperature change	Ice-water gravitational attraction
Juvenile water	Changes in geoid configuration

role, especially over the long term. The infilling of the ocean basins by sediment, for example, would tend to lead to a sea-level rise. With current rates of denudation Higgins (1965) has estimated that this could lead to a rise of 4 mm/100 years. This is equivalent to a rise of 40 m in a million years. Two very minor factors are the addition of juvenile water from the Earth's interior and the variation of water-level according to temperature. The latter could raise the level of the sea by about 60 cm for each 1 °C rise in temperature of the sea-water. The former could probably add about 1 m of water in a million years. The evaporation and desiccation of pluvial lakes, some of which had large dimensions (see p. 74), would be unimportant in affecting world sea-levels, adding a maximum of about 10 cm to the level of the sea were they all to be evaporated to dryness at the same time (Bloom, 1971).

Another cause of eustatic changes of sea-level, especially in the Holocene, is the process called 'isostatic decantation'. Isostatic uplift in the neighbour-hood of the Baltic basin and of Hudson Bay has led to a reduction in the volume of these seas, and the water from them has thus been decanted into the oceans to affect worldwide sea-levels. A comparison of the area and volume of the late-glacial precursor of Hudson Bay with Hudson Bay itself suggests that the volume of water decanted could only be sufficient to cause a rise in world sea-level of about 0·63 m. The contribution of the Baltic Sea would be even less. This factor can thus be largely ignored.

Thus these minor factors are of a very limited degree of importance in terms of a Quaternary time-scale, especially when they are compared with the changes brought about by glacio-eustasy.

Glacio-eustasy

During the first decades of the twentieth century, following on in part from the work of Suess, a number of workers, including De Lamothe, Deperet, Baulig, and Daly proposed that most sea-level oscillations and strandlines of the Quaternary were glacio-eustatic (see Guilcher, 1969). They believed, correctly, that sea-level oscillated in response to the quantity of water stored in ice caps during glaciations and deglaciations.

They proposed, and this proposal has been supported by some recent Australian workers, that there was a suite of characteristic levels in Morocco and elsewhere around the Mediterranean which could be related to different glacial events:

> Sicilian (80–100 m)
> Milazzian (55–60 m)—between the Günz and the Mindel
> Tyrrhenian (30–35 m)—between the Mindel and the Riss
> Monastirian (15–20 and 0–7 m)—between Riss and Würm
> Flandrian—the present post-Würm transgression.

The transgressions of the interglacial were succeeded by regressions during glacials, and the height of the various stages declined during the course of the Pleistocene (Fig. 6.1). Total melting of the two main current Ice Caps—

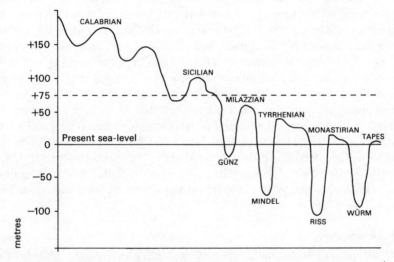

FIG. 6.1. The classic sequence of Pleistocene sea-levels, showing the downward trend in the elevation of raised beaches through time. The dashed line represents the approximate elevation of sea-level if the ice in Greenland and Antarctica were melted (after Frenzel, 1973, Fig. 92).

Greenland (2480 km³), and Antarctica (22 100 km³)—would raise sea-level a further 66 m if they both melted. Deep-sea core evidence, however, does not suggest that in previous interglacials of the Pleistocene these two Ice Caps did disappear, and without a general melting of them, sea-level would only have been a few m higher than now in the interglacials. This fact does not tie in too happily with the simple glacio-eustatic theory of progressive sea-level decline during the Pleistocene. Some factors other than glacio-eustasy must be responsible for the proposed high sea-levels of early Pleistocene times. Furthermore, because other factors have played a role, some local, few people now seriously believe that through height alone can one correlate shorelines over wide areas on the basis of a common interglacial age.

Nevertheless, low Quaternary sea-levels brought about by the ponding up of water in the ice caps were quantitatively extremely important. Donn *et al.* (1962), on the basis of theoretical considerations from known ice volumes reckon that in the Riss, possibly the most extensive of the glaciations, sea-levels might have been lowered by 137 to 159 m below current sea-level. In the Last Glacial (Würm–Wisconsin–Weichsel) they give a figure for lowering of rather less—105 to 123 m.

There is a growing body of geomorphological and sedimentary evidence to support this assessment. For example, the bottom of the Iroise (a body of water off Western Brittany) is covered down to 100 m below sea-level with solifluxion-derived materials which are only very slightly reworked by the action of the sea. Off New England, McMaster and Garrison (1966) claim changes down to 144 m as a result of detecting old shorelines. However, on the basis of isotopic dates for coral and associated material in the Great Barrier Reef (Australia), California, and SE. Caribbean sea areas, Veeh and Veevers (1970) favour the conclusion that 13 600 to 17 000 years ago, that is towards the end of the Last Glaciation, sea-level dropped universally to at least—175 m, some 45 m deeper than hitherto suspected.

The consequences of this low sea-level included the linking of Britain to the continent of Europe, the linking of Ireland to Britain (Whittow, 1970), of Australia to New Guinea, and of Japan to China (Emery *et al.*, 1971). The floors of the Red Sea (Olausson & Olsson, 1969) and the Persian Gulf (Saarnthein, 1972) were also dry land. The possible effects of such major changes in the geography of the earth on migrations of flora and fauna are discussed on p. 62.

Orogenic-eustasy

Although orogeny is normally regarded as being an essentially local factor of sea-level change, and eustasy as being of a world-wide nature, there is one class of process, here called orogenic-eustasy, whereby a local change can have world-wide effects. It therefore acts as some sort of a link between the two main types of change.

Figure 6.2 represents the sort of picture that one can envisage, and this is a situation that can easily be represented in the laboratory with simple materials. Two 'continents', represented by rectangular blocks of lead, float on a mantle of mercury. Water, representing the sea is poured on to the mercury so that the 'continents' are just submerged. One of the continents can then be deformed and a mountain created by the simple process of turning one part of a continent upright, thereby effectively halving the

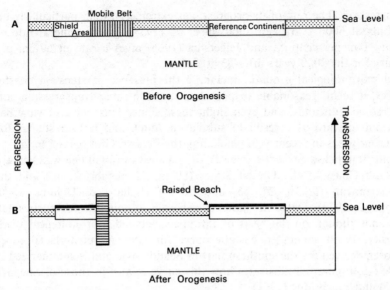

Fig. 6.2. The production of world-wide changes in sea-level as a result of orogenesis (after Grasty, 1967).

width and doubling the thickness of the 'mobile belt'. The deformed 'continent' will displace the same amount of mercury as the undeformed continent, although the mountain will have a deeper root. Thus the level of the mercury will remain the same, but the water now has a larger area to spread over, so that it will spread out reducing the depth of the water. The stable 'continent' will thus emerge from the sea. In effect one is producing a world-wide regression of the sea by means of a local orogenic event (Grasty, 1967). It has been calculated that an increase of only 1 per cent in the area of the oceans would lower the sea-level by about 40 m, assuming the average depth of the oceans to be 4 km. Over a long time-period this process could be significant, though it probably cannot explain the shorter amplitude sea-level fluctuations of the Pleistocene. On the other hand, the gradual fall of interglacial sea-levels during this epoch which has frequently been proposed (see p. 171) could have been caused partly by this mechanism.

Bloom (1971, p. 355), on the basis of recent information derived from studies of global tectonics, estimated that as ocean basins are spreading at rates of up to 16 cm per year, 'The spreading of the ocean basins since the Last Interglacial could accommodate about six per cent of the returned meltwater, and the post-glacial shorelines would be almost 8 metres lower than the interglacial shorelines of 100 000 years ago'.

The post-Glacial rise in sea-level or Flandrian transgression

Round about 14 000 years ago the lowering of sea-level was succeeded, as the ice melted, by a large rise in sea-level: the Holocene or Flandrian transgression. The rate of its progress was extremely rapid, particularly up until about 6000 years ago. Godwin *et al.* (1958), give a general rate of around 1 m per century, and Jelgersma (1966) gives a rate of 60 cm per century for the 2000 years after 8300 B.P.

On morphological grounds, including the presence of steps on coastal shelves, it seems reasonable to postulate that the rapid transgression suffered some stillstands, and even slight regressions. This point of view has been put forward as a result of submarine notch and terrace studies for numerous areas in recent years including the Persian Gulf (61–64 m, 40–53 m, and 30 m below present sea-level), the Bass Strait (60 m below sea-level), the Gulf Coast of the United States (60 m, 32 m, and 20 m) and the Mediterranean (5, 10, 27, 55, and 96 m) (Ballard & Uchupi, 1970; Flemming, 1972). Dates are not yet generally obtainable for such small stillstands though on the basis of extremely detailed morphological work (Mörner, 1969) a detailed eustatic curve for late and post-glacial times has been derived for the southern part of Scandinavia and is summarized in Table 6.2. Figure 6.3 illustrates how this curve relates to those of Shepard (1963) and Fairbridge (1961).

Whether such minor stillstands took place or not, the general Holocene trend until around 6000 B.P. was one of rapidly rising sea-levels. Over low-

Table 6.2
Eustatic pattern for the Holocene

Date (B.C.)	State of s.l.	Height of s.l. relative to the present (m)
11 750–10 700	Rising	−63 to −56
10 700–10 350 (Agård Interstadial)	Rising fast	147
10 350–10 300 (Fjaras Stadial)	Fall	−49
10 300–9 950 (Bølling Interstadial)	Rise rapidly	−42
9 950–9 800 (Older Dryas)	Fall	−46
9 500 (Early Allerød Interstadial)	Rapid rise	−42
9 500–9 000	Fall	−45
9 000–8 050 (Younger Dryas)	Slow rise	−42
7 800	Rapid rise	−36
7 800–7 700	Fall	−38·5
7 700–7 330	Almost stationary	−38
7 300–6 900/6 800	Very rapid rise	−18·5
6 900–6 800 to 6 500–5 800	Falling	−20·3
6 500–5 800	Stationary	
5 800–5 000	Rapid rise	−10
4 500	Slow rise	−6·6
4 300		−5·2
4 200–3 950		−6
3 900		−3·55
1 650–1 400		+0·4

(After Morner, 1969)

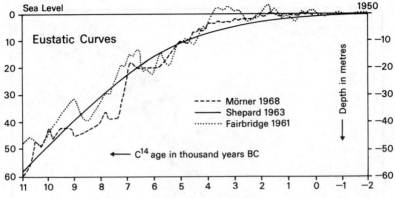

FIG. 6.3. Holocene eustatic curves (after Mörner, 1969, Fig. 160).

angle shelves this rapid rate of rise means that the sea must have advanced laterally at a quick rate. In the Persian Gulf region, for instance, there was a shoreline displacement of around 500 km in only 4000–5000 years, a rate of no less than 100–120 m per year. This must have had profoundly disturbing results for inhabitants of the coastal plains. Even in southern Britain, off south-east Devon, it has been calculated that between 9000 and 7000 years B.P. the sea was rising at a rate of about 1·5 m/100 yrs, which would mean a lateral coastline migration in that area of about 8 m per year (Clarke, 1970).

Figure 6.4 illustrates the way in which the shorelines of 11 000 and 15 000 years ago compare to those of North America today. The shoreline 15 000 years ago, being at a lower level, was a considerable distance across the continental shelf, and the various groups of islands off the east coast of Florida were linked up to form much larger land areas.

The main problem that arises with the interpretation of this Holocene transgression lies in what happened after about 6000 B.P. There are three fairly distinct schools of thought on this, though it has to be stated that the arguments are about possible changes of the order of only a couple of metres (Table 6.3). It is generally accepted that in the last six millennia the rate of sea-level rise, if present, has been far less than it was in the Early Holocene. One point of view has it that there has been a continuously rising sea-level to the present time, though the rate of rise has diminished with time (Shepard's hypothesis (1963)). Godwin *et al.* (1958) on the other hand hypothesize that sea-level rose steadily until about 3600 B.P. whereupon it has more or less remained constant. Fairbridge (1958) and others (Mörner, 1971) have maintained, in contrast to these other two ideas, that Late Holocene sea-level oscillated to positions both above and below the present level. He suggested that the sea was at levels 1–4 m above the present about 5 times between 6000 B.P. and the Middle Ages (Table 6.4). A considerable amount of evidence has been raised against Fairbridge's concept. Some people have reworked or reinvestigated some of the sites claimed by

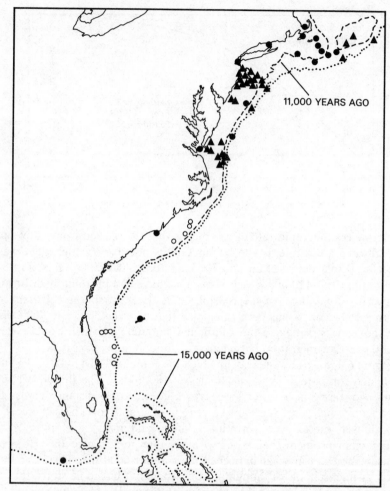

Fig. 6.4. A comparison of the Atlantic coast shoreline of the United States at 15 000 years ago, 11 000 years ago, and at the present. Confirmation that the continental shelf was once laid bare is found in discoveries of elephant teeth (triangles), fresh water peat (dots), and the shallow-water formations called oolites (circles) (after Emery, 1969).

Fairbridge to illustrate high Holocene stillstands and they give a Pleistocene rather than Holocene age for the raised beaches and terraces. Jelgersma (1966) has said that if the high sea-levels had taken place one would expect that coastal plains would have been inundated on a very large scale. She says that data from the Gulf of Mexico, Florida, and the Netherlands fail to reveal such a degree of transgression. Very detailed archaeological researches in relatively stable parts of the Mediterranean, using diving techniques, have convinced Flemming (1969) that to within an accuracy of ± 0.5 m there has been no eustatic change of sea-level in the Mediterranean in the

Table 6.3

The stages of world sea-level change since 13 000 B.P. according to different sources

Years B.P.	Shepard (1963)	Schofield (1960)	Fairbridge (1961)	Godwin et al. (1958)
		Sources		
1 000	−0·5	+1	+1	
2 000	−1	+2	−2	
3 000	−2	+3	−3	
4 000	−3	+5	+2	
5 000	−4	−2	+3	0
6 000	−7	−0·5	0	−4
7 000	−10	−4	−6	−9
8 000	−16	−19	−16	−17
9 000	−22	−33	−14	−28
10 000	−31	−36	−32	−35
11 000	−40			−44
12 000	−48			−52
13 000	−58			−62

(Levels in m)

last 2000 years. Dating of freshwater peats in Australia, one of Fairbridge's field areas, by Thom *et al.* (1969) fails to indicate that sea-level rose above its present position between 2985 B.P. and 9000 B.P. Likewise dating of Chenier ridges in Queensland leads to a broadly comparable conclusion (Cook & Polach, 1973). Similarly, after an expedition around some Pacific atolls, Newell and Bloom (1970) said 'We found no unequivocal evidence for recent higher sea-level, and abundant evidence that the characteristic morphology of the Indo-Pacific coral reefs is most probably in adjustment to the slow rise in sea-level that has characterized the last 6000 years.'

One of the main lines of evidence that has been used to substantiate the high Holocene sea-level concept is the presence of small raised terraces

Table 6.4

Fairbridge's Holocene oscillatory sequence

Transgression	Emergence	sea-level (m)	Date (B.P.)
Older Peron		−3 or 4	5 000
	Bahama	−3	4 300
Younger Peron		+3	3 900–3 400
	Crane Key	−2	3 300
	Pelham Bay	−3	2 400–2 800
Abrolhos		+1½–2	2 300
	Florida	−3	2 000
Rottnest		+½ to 1	1 200–1 000
	Paria		700

(Fairbridge, 1958)

('Daly levels') in many parts of the tropics. Radiocarbon dates for slightly elevated reefs of Holocene age, which cluster around about 4000 years B.P., as reviewed by Stoddart (1969), do indicate the possibility of a slight Holocene transgression, but he says, there are so few, that they could be due to local emergence.

However, many of the terraces, from which samples of *in situ* coral have been dated by various isotopic means, indicate that a large number of the 'Daly Levels' may be the product of earlier high stands of sea-level. Higher sea-levels than the present have been suggested for the middle of the Würm (around 30 000 B.P), and for pre-Würmian times, notably 70 000 to 180 000 B.P, and to a lesser extent 190 000–240 000 B.P. In general, as explained on p. 181, it seems probable that sea-level stopped at or slightly above its present level for much of the period 70–190 × 10³ years ago. This would give a much longer time for the terraces to develop than the limited number of years available during supposed Holocene oscillations. Recent work on rates of erosion on coral islands suggests that time would have been inadequate in the Holocene to produce the degree of planation recognized (Stoddart, 1971).

There is certain evidence from present tide-gauge records to suggest that following on from the current post-neoglacial (Little Ice Age) amelioration in climate there has been a corresponding rise of sea-level. It is of course difficult to separate current tectonic submergences and other factors from the eustatic effect of current glacial melting, but there is a certain consistency in the records which leads one to think that the eustatic effect may be important. Some of the data are summarized in Table 6.5. It is interesting from

Table 6.5

Current rates of sea-level rise

Source	Rate (cm/100 yr)	Date
Gutenberg (1941)	19·4	1880–1942
Fairbridge and Krebs (1962)	12·0	1900–50
Fairbridge and Krebs (1962)	55·0	1946–56
Scholl (1964)	18	1914–64
Scholl (1964)	12	1940–64
Donn and Shaw (1963)	42	1890–1940
Donn and Shaw (1963)	24	1940–60
Hawkins (1971)	25	1916–62

the glacio-eustatic viewpoint that the tendency towards a reduced rate of glacial retreat and of a reduction or reversal in the climatic amelioration of the twentieth century (see p. 137) has, according to the studies of Scholl (1964), and Donn and Shaw (1963), led to a reduction in the rate of eustatic rise in the last two or three decades. Similarly, Binns (1972) claims to have found evidence of a number of shorelines related to a fall in sea-level during the cold neo-glacial phase (see p. 116) around 2500–2400 years B.P.

In some localities present-day subsidence, combined with eustatic rise is of a sufficient magnitude to present a threat to low-lying settlement concentrations. In the case of London, for instance, historical records show that high tide levels and surge levels relative to Newlyn Ordnance Datum are becoming progressively higher (Fig. 6.5), with an increase of the order of 1·3 m between 1791 and 1953. This increases flood risk, but it is not clear as to how much this is due to subsidence and eustatic rise alone, and to what extent embanking by man, changes in water temperatures affecting tides through changing viscosity of the water, and changes in climatic conditions

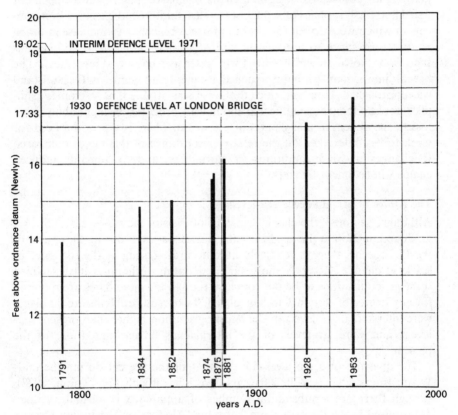

FIG. 6.5. Increasing high tide levels at London Bridge (after Horner, 1972, Fig. 1).

(including rainfall and wind directions), have played a significant role (Bowen, 1972; Horner, 1972).

A comparable picture is evident at Venice where storm surges (locally called 'acqua alta') have reached new levels of intensity and frequency over the last few years (Berghinz, 1976). Out of 58 surges recorded in the past hundred years, 48 occurred in the last 35 years, and 30 in the last 10 years. In other words in the first 65 years there was one acqua alta every 5 years, in the following 25 years almost one per year, and in the last ten years three per year. Eustasy plays a role, but subsidence appears to be the major problem, with the rate increasing fivefold from 0·9 mm/year for the period 1908–25, to 5mm/year for the period 1953–61.

Elsewhere in the world, the relative stability of sea-level in the last few thousand years may be one contributing factor to the widespread sand loss from many beaches, with consequent erosion and threats to human activities (Russell, 1967). As long as the Flandrian transgression was taking place at an appreciable rate, new areas of coastal plain were being inundated and fresh supplies of sediment were encountered, the coarser components of which were driven forward to produce beaches. An increase in beach volume took place so long as the rise of sea-level continued, and surplus sand was blown downwind to form extensive coastal dune tracts. The beach-dune system probably reached its greatest volume as the stillstand was approached. However, once that level was attained, new sediment supplies were no longer encountered, and marine processes brought about a net loss to the system in many parts of the world. There is a pressing need for work to be undertaken on the relative importance of this factor compared to changes in wind conditions or storminess, and to humanly induced changes in sediment budgets.

The nature of pre-Holocene sea-levels

Although, as noted, the classic sequence of Pleistocene sea-levels involves a gradual reduction in the height of interglacial sea-levels during the course of the Pleistocene, there is relatively little accurate dating of the sequence of sea-level change. Recently, however, new dating techniques have enabled some generalizations to be put forward, though far more dates are required for any degree of certainty to be reached. Two main problems are involved: were interstadial sea-levels of the Würm glaciation higher than present sea-levels, and were sea-levels of the interglacials higher than those of the present?

The question of a sea-level close to that of the present during the mid-Würm Interstadial, say 30 000 years ago, is a difficult one (Thom, 1973), though there are a substantial number of apparently confirmatory dates from raised beach deposits from Tanzania, Aldabra (in the Indian Ocean), the Atlantic, and the Red Sea. It needs to be pointed out, however, that dates of this order are close to the limit of C14 dating, and even slight contamination could produce misleading dates. Indeed, for Red Sea and Aldabran samples discrepancies have been revealed between C14 dates and those obtained by other radiometric methods of the Uranium Series (Thomson and Walton, 1972). The difficulty of radiocarbon dating of shells and corals of this age can be highlighted by considering the way in which contamination by a certain percentage of modern calcium carbonate can affect the apparent C14 age of a carbonate which is in reality 100 000 years old. The following situation has been found, and indicates that slight contamination of a sample of interglacial age could appear to give a date of the

Per cent Contamination by modern CaCO₃	Apparent C14 years
50	5 600
10	19 000
5	24 500
1	37 000
0·01	74 000

Interstadial of the mid-Würm (Newell, 1961). As with the Holocene sea-levels, non-eustatic factors may have played a role in some situations, so that this proposed interstadial high stand of sea-level is at present only a possibility. Nevertheless, if a mid-Würm Interstadial was a reality as seems likely (see p. 53) it would be expected to have caused a relative sea-level rise, though on the basis of such climatic information as is available one might expect eustatically controlled sea-levels during these relatively warm interstadials to be at −40 to −50 m. Temperature conditions in the interstadials were only sufficient to cause partial melting of the big ice caps. The Laurentide and Scandinavian Ice Caps persisted during the last Glacial, colder-than-present vegetation was present in Western Europe, and oxygen-isotope records from deep-sea and Greenland ice cores show values not equivalent to interglacial or present conditions. The pre-Würm sea-level situation is perhaps a little clearer, though the accuracy and frequency of Uranium Series dates still leaves much to be desired. In gross terms it is possible to see from Table 6.6 that between around 70 000 yrs. B.P. and around 200 000 B.P., sea-levels were relatively if not absolutely high. Certain suggestions have been made that it is possible to see certain more limited phases of high stillstands within this wide period, and dates of 200 000 ± 20 000, 120 000 ± 12 000, and 80 000 ± 8000 have been suggested for three main levels. The earlier of these levels, probably belong to the Eemian (Saalian–Weichselian) Interglacial, and have in the past been given a variety of names including Ouljian, Normannian, Eu- and Neotyrrhenian, Sangamon, and Karimbolian.

Some of the pre-Würm dates for high sea-levels appear to correlate closely with both maxima on theoretical insolation curves (Veeh and Chappell, 1970) (see p. 210), and with temperature maxima established from deep-sea cores (Broecker et al., 1968). The multiple nature of some of the raised beaches (for example, the Upper and Lower Normannian, and the Eu- and Neotyrrenhian) may be related to the several pre-Würm, high sea-level stands that the Uranium Series dates are beginning to suggest. On

Table 6.6

Radiometric dates for pre-Holocene sea-levels

(*Yrs.* B.P.)

Source	Location	Dates of transgressions
Stearns and Thurber (1967)	Morocco	95 000–60 000
		200 000–115 000
		250 000
Osmond *et al.* (1965)	Florida	130 000
Ku (1968)	Barbados	80 000
		105 000
		120 000
Broecker and Thurber (1965)	Bahamas and Florida	85 000
		130 000
		190 000
Veeh and Chappell (1970)	New Guinea	50 000–30 000
		74 000
		118 000–140 000
		180 000–190 000
Broecker *et al.* (1968)	Barbados	82 000
		103 000
		122 000
Mesolella *et al.* (1969)	Barbados	82 000
		105 000
		125 000
		170 000–230 000
		250 000
Milliman and Emery (1968)	Atlantic area	30 000–35 000
Thurber *et al.* (1965)	Eniwetok	120 000
Stoddart (1969)	Aldabra	30 000
	Tanzania	26 750–34 400
Veeh and Giegengack (1970)	Red Sea	90 000
Valentine and Veeh (1969)	California	120 000
Veeh and Valentine (1967)	California	130 000–140 000
Thomson and Walton (1972)	Aldabra	127 000
Birkeland (1972)	California	104 000
		131 000
		200 000+

stratigraphic grounds, however, it needs to be pointed out that certain British raised beaches have been assigned not to the Eemian but to the last interglacial but one (the Holsteinian or Mindel/Riss) (Guilcher, 1969).

It is not possible to say anything with any great certainty about world sea-levels prior to about a quarter of a million years ago. The sedimentological and geomorphic evidence becomes even more fragmentary, and dating procedures become more difficult. The evidence that ice caps and glaciers were established and changed their volumes during the course of the Tertiary would suggest that world sea-levels probably changed as a result of glacio-eustatic changes (see Tanner, 1968) in early and pre-Pleistocene times. In addition, orogenic-eustasy connected with Cainozoic mountain building would probably have been important, probably causing a world-wide drop in sea-levels. In various parts of the world there is evidence for early high sea-levels (Andrews, 1975). In Australia and New Zealand sea-level may have been about 180 m above present level during the Middle Pliocene (6 million years ago). In South Carolina (U.S.A.) sea-level has fallen 76 m since Miocene times, and in the eastern U.S.A. and New Zealand sea-level at the Pliocene–Pleistocene boundary was about 30 m above present. In many parts of southern England there is morphological evidence for an Early Pleistocene marine platform at about 210 m, and a similar bench may exist in Wales.

However, over the time-span involved here it is difficult to believe that many parts of the world have been totally stable, especially in view of the on-going changes in earth geometry associated with the plate tectonics model. Thus attempts to develop a world wide curve of sea-level for the Early Pleistocene and periods before that are likely to fail. Even within the Late Pleistocene local movements have been extremely important in many areas. The various mechanisms involved in these local changes will now be discussed, and this will be followed by a regional examination of some of the ways in which eustatic and local forces have combined in given areas to produce complex pictures of emergence and submergence.

Isostasy

The earth's crust responds when a load is applied to or removed from it. Thus during the course of a cycle of denudation as perceived under the Davisian model, erosion would remove a considerable mass of rock from on top of underlying strata and a certain amount of compensatory uplift would delay the point at which base level would be attained. However, it is unlikely that this isostatic effect would be of more than local significance during the relatively short time-span of the Pleistocene and Holocene.

However, isostasy would be important in two main ways: by the application and removal of large masses of ice to certain parts of the earth's crust; and by the application and removal of large bodies of sea-water, and, occasionally, lake water, from the continental shelves, and from lake basins.

The latter mechanism has recently been called *hydroisostasy* and is probably of less importance than *glacio-isostasy*, but its effects have often been forgotten. The hydroisostasy theory can be summarized thus. Eustatic sea-level changes brought about by the melting and freezing of the ice caps would alternately add and take away water from the ocean basins. This water would thereby add or remove a load from the ocean floors, and if one assumes that the density in the subcrustal zone is nowhere less than 3·00 or 4·00 then total isostatic adjustments to the water loads resulting in depression of sea-bed would be expected to range from one-third to one-quarter of the effective depth of the water (Higgins, 1969). In fact, however, the rate and amount of hydroisostatic deformation would vary from place to place according to various factors. Coasts with nearby ocean water more than 100 m deep had the load of water from the post-glacial eustatic rise of sea-level added early and close to the shore, whilst coasts which now border shallow seas had the load added late and, generally, far offshore. One would expect that the amount of submergence would be roughly proportional to the proximity of deep water. This has tentatively been confirmed by recent studies on the coast of America. In this interpretation, it is the shallow offshore depth of the Everglades area of Florida which causes that region to be relatively stable (Fig. 6.6) in comparison with New England (Bloom, 1967).

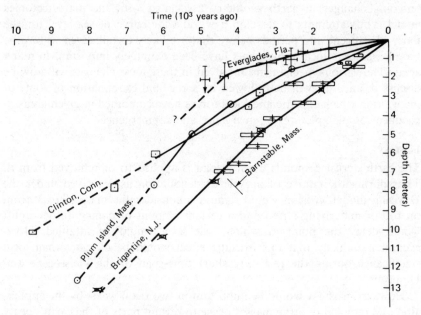

FIG. 6.6. Comparison of published eustatic sea-level curves from the eastern seaboard of the United States to illustrate the effects of hydro-isostasy (after Bloom, 1967).

Other factors which might affect the degree of submergence consequent on the sea-level rise would be the local sub-crustal density, its dynamic viscosity, and the degree of isostatic adjustment achieved before loading or unloading began. Estimation of the general effects of this mechanism, however, suggests that melting of all the present Antarctic ice would raise sea-level eustatically by around 60 m, but that compensatory hydroisostatic sinking of the ocean floors would reduce the effective sea-level rise to about 40 m, that is to around two-thirds.

The effects of hydroisostasy, however, can also be seen in the case of pluvial Lake Bonneville, which, as already stated, attained a great depth and area during the pluvial phases of the Pleistocene. It attained a depth of some 335 m. The shorelines around it, which are well known following the studies of G. K. Gilbert and later workers, are all warped according to their position relative to the area of maximum water depth. The Bonneville high shoreline is at least 64 m higher on islands in the ancient lake than it is at the periphery (Crittenden, 1963). Pluvial Lake Lahontan also shows warping, but only of 6 to 9 m. More recent evidence of this phenomenon can be seen at the man-made Lake Mead in the U.S.A., where the building of the Hoover Dam has led to the impounding of 40 000 million tons of water over 600 km², creating a usual load of about 140 pounds per square inch. This has caused subsidence of the order of 170 mm over the centre of the lake between 1935 and 1950. The degree of subsidence decreases progressively from the lake centre (Longwell, 1960).

The principle of glacio-isostasy can be summarized thus: during glacial phases water loads were transferred from the oceanic 70 per cent of the earth's surface to the glaciated 5 per cent. This led to depression of the crust, whilst the release of the weight of the ice resulting from melting leads to uplift (Fig. 6.7).

Relatively little is known about the nature of the depression sequence, but as the emergence sequence is both evident in raised beaches and measurable at the present time, more is known about this (Andrews, 1970). It is possible to identify three main stages of isostatic response: *the period of restrained rebound* as the ice sheet begins to lose mass; *the period of post-glacial uplift*

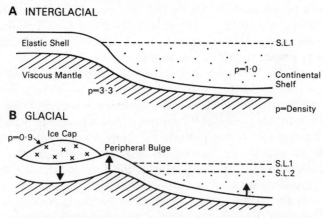

FIG. 6.7. A simplified, model of the effects of glaciation on local sea-levels.
   A represents the interglacial or pre-glacial situation.
   B represents the situation when an ice sheet has developed. Depression takes place beneath the ice sheet as a result of the transference of mass from the oceans, but the land rises as a bulge at some distance from the ice cap. The continental shelf rises as a result of the removal of a mass of water as sea-level falls.

during which ice loss takes place at an accelerating rate giving a smooth acceleration of uplift; and, *the period of residual glacio-eustatic recovery* when some coastal uplift still goes on in spite of total ice removal. In most areas this is the position that we are in today. Because of this sequence the gradients of tilt lessen on younger and younger shorelines in a way that is clearly related to the exponential form of post-glacial isostatic uplift. Some quantitative data are available from Scotland which suggest that in the eastern part of the country a degree of tilt of 18 mm/km/1000 years was established between 9500 and 5500 B.P., but that this rate decreases to a tilt of 10·9 mm/km/1000 yr between 5000 B.P. and the present.

In areas peripheral to the ice sheets, such as parts of the eastern coast of the United States, the Baltic, and North Seas, there are indications that zones not subject to an ice load bulged up during glacial phases (Newman *et al.*, 1971), perhaps because of volumetric displacement of the upper mantle's low velocity layer, but that they have collapsed peripherally in post-glacial times, giving greater submergence than could be explained by the eustatic Flandrian (Holocene) transgression alone. This is thought to be the result of some compensatory transference of sub-crustal material.

In the areas that were covered by ice, however, the degree of maximum isostatic uplift has been considerable: around 300 m in North America and 307 m in Fennoscandia, but less in Great Britain (Fig. 6.8, a, b, and c). The Ice Caps of Greenland and Antarctica, moreover, are still exerting sufficient weight to create a considerable degree of isostatic depression (see Fig. 6.8d). Much of the bedrock surface of interior northern Greenland is currently at or below present sea-level. Gravity readings and ice thickness determinations obtained by trans-Greenland expeditions suggest that before the ice sheet formed the presently low-lying bedrock areas of northern Greenland, some of which extend below sea-level, were in the form of a plateau about 1000 m above sea-level. If the ice were to be removed, the bedrock surface would slowly rise up to this height once again.

The effect of glacio-isostasy on the lakes of Finland is also of interest. It is known by tradition that the woodland and meadows on the low southern shores of the Great Finnish lakes had a tendency to become more swampy and marshy with time, whereas the northern shores tended to become drier. This is a consequence of the post-glacial tilting which has sometimes had spectacular results: fjords opening towards the north, to the Gulf of Bothnia, have been converted into lakes by the development of a threshold. Equally, many of the large lakes used to have their outlets on their northern sides, but they too have had them diverted to the south shores by the same process. The rivers draining the lakes of the southern sides have not had time to fully adjust their profiles to this phenomenon and as a result they show rapids and falls which have proved to be popular sites for electricity development in this century. Elsewhere some of the lakes have dried up through decantation. In the north of the country the uplift of the land has

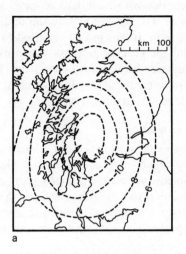

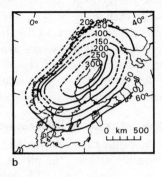

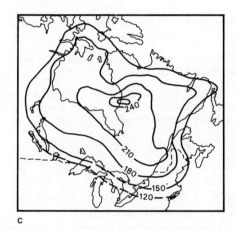

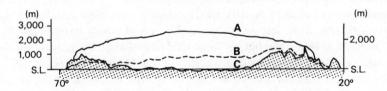

FIG. 6.8. The effects of glacio-isostasy on different land areas.

    (a) Generalized isobases for the main post-glacial raised shoreline in Scotland (m)

    (b) The degree of isostatic recovery (m) of Scandinavia in the last 10 000 years

    (c) Maximum post-glacial rebound of north-eastern North America (m)

    (d) Cross-section of north Greenland showing the present shape of the ice sheet (A), the present level of the bedrock surface, (C), and the estimated profile of the landmass with no ice load (after Hamilton, 1958).

also caused difficulties for port authorities who have had to deal with a progressive shallowing of their harbours. On the bonus side the uplift has provided further usable land for the nation. It was in fact the need to allot newly emerged land to owners around the Gulf of Bothnia which led the director of the cadastral survey of Finland, Efraim Otto Runeberg (1722–70), to postulate in 1765 that small movements of the Earth's crust were responsible for many of the gains in land (Wegmann, 1969).

Miscellaneous causes of local changes in level

One of the prime causes of the observed changes of the land relative to the sea is orogenic activity, the process by which mountains are built. Signs of Pleistocene vulcanity and earth movements are visible in many parts of the world (Table 6.7, Fig. 6.9), and Charlesworth (1957) has written that the Pleistocene 'witnessed earth-movements of a considerable, even catastrophic scale. ... The Pleistocene indeed represents one of the crescendi in the

Table 6.7
Areas of Pleistocene vulcanicity and tectonic uplift

Europe
  Greece, the Aegean, Vesuvius, Etna, Sardinia, Catalonia, Massif Central, Northern Bohemia, Romania, Silesia, Eifel area, Spitzbergen, Iceland.
Asia
  Armenia, Asia Minor, Caucasus, Iraq, North Palestine, Transjordan, Arabia, Dead Sea and Galilee, Northern Siberia, Mongolia, Manchuria, Korea, China, Sea of Okhotsk, Japan, Kuriles, Java, Sumatra, and various Pacific islands.
America
  Alaska, Sierra Nevada, West Indies, Central America, the Andes.

(After Charlesworth, 1957)

earth's tectonic history.' This is not, however, a universally accepted point of view. In place of this 'crescendo' belief, others would maintain that the Quaternary has witnessed mountain building on a scale that is not dissimilar from that of previous eras, whilst others believe that the Quaternary was a new and distinct phase of activity which replaced supposed stability of the Middle Tertiary. This last hypothesis, the neo-tectonics hypothesis, has much support at this time, but all three models involve a recognition of the considerable extent of orogenic movements in the Quaternary (King, 1965).

The zones where orogenic activity has been most intense in the Pleistocene have been at the margins of the various tectonic plates that have been recognized over the last fifteen years. Seismic activity, vulcanism, and mountain building, occur for example, in a well-defined series of narrow belts (Fig. 6.9), with that surrounding the Pacific Ocean being especially notable. On the other hand, some areas, the continental platforms, located away from plate margins, have suffered from relatively little mountain building during the Pleistocene. They stand in contrast to the areas of new fold mountains, some of which may have been uplifted as much as 2000 m in the last few million years.

In some localities, and on a much more limited scale, the flattening of sediments by the weight of overlying material (autocompaction) can lead to subsidence of some consequence. It is often apparent in peat and other such materials which have a very high porosity and a weak skeletal framework of vegetable fibres. Salt marsh peats, for example, which make up a large part of many transgressive sedimentary deposits (see p. 198), have an 80 per cent porosity, and frequently in section one sees logs that have become flattened from their original round shape to a more oval form (Kaye and Barghoorn, 1964). The subsidence caused by compaction of Holocene beds in Holland is estimated at 2·5 cm/100 years (Veenstra, 1970).

In some localities man may have caused some fall in land levels relative to the sea. One of the clearest indications of this is provided by the example of Venice. Currently there is an increasing flood risk which is leading to the frequent inundation of St. Mark's Square and other parts of the city. Although currently rising sea-level of eustatic origin and more long continued subsidence in the area both play a role, one of the prime causes of the problem is the abstraction of groundwater by large new industrial complexes on the other side of the Venetian lagoon. This abstraction has caused subsidence to take place (see Fontes and Bortolami, 1973, and Gambolati *et al.*, 1974).

Operating on a broader scale than orogenic movements are epeirogenic movements. These do not involve complicated deformation with folding, faulting, tilting, and warping, but involve large elevations and depressions of continents and ocean basins, with warping restricted to a so-called marginal hinge line. On the landward side there would tend to be elevation, producing stairways of terraces, many of which have been attributed to eustasy, whereas there would tend to be subsidence on the seaward side. Broad-scale epeirogenic movements have marked the evolution of the African continent, where broad planation surfaces initiated during phases of tectonic stability are separated by massive scarps initiated during each uplift episode (King, 1962).

Other broad-scale changes in local sea-levels might be caused by changes in the geoidal configuration (Mörner, 1976), but as yet little firm information is available on the likely effects of this mechanism during the Pleistocene.

Clark (1976) has postulated another possible mechanism of local sea-level change: that the growth of massive Pleistocene ice sheets, such as that of Canada, would cause sufficient gravitational attraction for sea-level to rise locally relative to the land. As the ice sheets melted and lost mass, the sea-level would also fall in response to the reduced gravitational attraction. He suggests that this ice-water gravitational effect alone could cause raised beaches to occur 85 m above present sea-level in Hudson Bay.

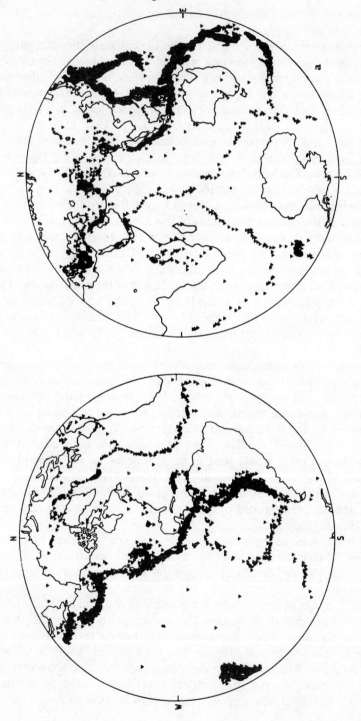

FIG. 6.9. The distribution of current tectonic activity in the world.
A. The distribution of earthquakes between Jan. 1965 and Dec. 1967 (after Tarling and Tarling, 1971, Fig. 40, pp. 94 and 95).

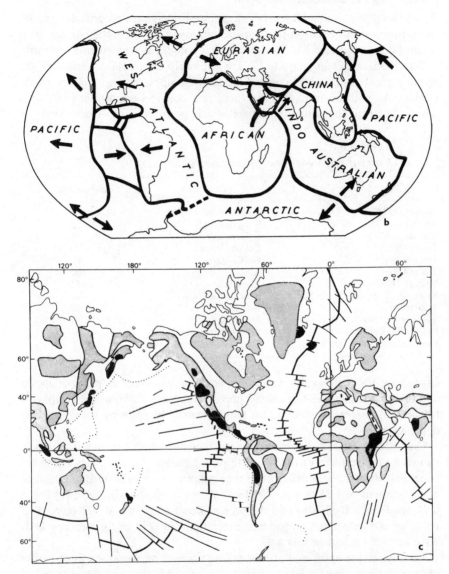

FIG. 6.9. (cont.) B. Suggested plates of the Earth's crust (after Tarling and Tarling, 1971, Fig. 40b, p. 95).
    C. The major tectonic features of the world. Light shading = continental platforms; irregular shading = continental shields; dark shading = Tertiary folded mountain chains; black areas = Cainozoic volcanic regions; dotted lines = oceanic trenches; heavy lines = active rift systems of oceanic ridges; light lines = oceanic faults (after P. J. Wyllie, in Smith, 1973, p. 9, Fig. 1.3.6).

Current rates of subsidence and uplift

Precise levelling, archaeological evidence, and tide-gauge records are among the sources of information that can be employed to assess the current rates of subsidence (Table 6.8) and uplift which complicate the eustatic picture.

The main areas of subsidence are probably the deltaic basins of the world's great rivers such as the Rhine, Mississippi, Rhone, and Narbada (India). As with eustatic changes there is a considerable quantity of data which illustrate the order of change which is progressing. Other areas of

Table 6.8

Current or recent rates of subsidence

Source	Location	Rate cm/100 yr.
Veenstra (1970)	NW. Germany	2·5
Veenstra (1970)	S. Netherlands	10–20
Veenstra (1970)	S. Denmark	15
Tjia (1970)	Tokyo (Japan)	10
Tjia (1970)	Osaka (Japan)	12
Tjia (1970)	Alaska	100
Vasil ev (1969)	Black Sea	30–52·5
Kvitkovic, J. & Vanko (1971)	New Carpathians	up to 5
Tjia (1970)	Indonesia	30
Van Veen (1954)	Netherlands	25 (since 7 200 B.P.)
Churchill (1969)	East Anglia & Kent	9
Coleman and Smith (1964)	South-central Louisiana	7·3

change are the tectonically unstable volcanic areas, such as parts of Japan and the East Indies. Also, as noted previously, parts of the world peripheral to zones of current isostatic uplift may currently be showing some isostatic depression.

One effect of coastal subsidence and landward uplift in deltaic areas is shown by the terraces of the Mississippi in Louisiana and Texas. These are progressively steeper in gradient as they become older. The Williana terrace has a gradient of 1·86 m/km, the Bentley 0·66–1·49, the Montgomery 0·47–0·94, the Prairie 0·20–·045 and the present flood-plain only 0·002–0·26. As a result of the associated subsidence more than 3000 m of Quaternary beds have been deposited in this area.

Locally, major uplift may still be taking place at measurable rates because of earthquake and related activity. Some of the measured vertical movements, determined by accurate geodetic levelling and other methods have been quite considerable, as Table 6.9 shows. The movements take place both because of individual seismic events and because of a more general process of gentle seismic creep.

The degree of isostatic adjustment taking place at the present time has also been estimated by analysis of tide-gauge data. Trends can be identified in rates, with minimum rates occurring away from the former centres of ice cap

Table 6.9

Current rates of vertical movement associated with seismic activity

(a) *Specific events*

Source	Location	Displacement	Date
Plafker (1965)	Alaska	10 m–15 m	1964
Plafker (1965)	Yakutat Bay (Alaska)	14·3 m	1899
Twidale (1971)	New Zealand (Murchison quake)	5 m	1929
Twidale (1971)	Adelaide (Australia)	5–8 cm	1954
Daly (1926)	California	7·01 m	1872
Daly (1926)	Sonora (Mexico)	6·10 m	1887
Daly (1926)	Japan	6·10 m	1891
Daly (1926)	India	10·67 m	1897
Daly (1926)	California	0·91 m	1906
Daly (1926)	Formosa	1·83 m	1906
Daly (1926)	Mexico	0·61 m	1912
Daly (1926)	Nevada	4·57 m	1915

(b) *Gradual seismic movements* (rates, m/1000 yrs.)

Source	Location	Rate
Kvitovic and Vanko (1971)	Carpathians	up to 1
Kafri (1969)	European U.S.S.R.	up to 14·6
Kafri (1969)	Israel (Yoqneam–Nazareth traverse)	60
Tjia (1970)	New Zealand	4·0
Tjia (1970)	Indonesia	0·2–1·5
Collins and Frazer (1971)	New Zealand	11–12
Collins and Frazer (1971)	Japan	0·7–1·6
Collins and Frazer (1971)	Garlock Fault (California)	10
Schumm (1963)	California	3·96–12·8
Bandy and Maricovich (1973)	California	5–8

growth. In Finland, for example, during the present century, rates in the south at stations like Helsinki and Hamina have been about four times lower than those at the head of the Gulf of Bothnia (Table 6.10) (Oulu, Kemi, etc.).

Post-glacial sea-level changes in northern Europe: the combined effects of eustasy and isostasy

Thus far we have been concerned solely with the operation of the various individual factors that have influenced sea-level changes in the Quaternary Era. In reality, of course, the sea-level fluctuations at any one point involve a combination of several factors. This is especially clearly illustrated in the case of the countries of north-western Europe, where both glacio-eustasy on a world-wide basis and glacio-isostasy on a more local basis have acted together in post-glacial times to give the observed sequences of sea-level

Table 6.10

Current rates of isostatic uplift as determined for Finland from tide-gauge data

Station	Mean rate (cm/10 yr.)
Kemi	7·37
Oulu	6·53
Raahe	7·63
Pietarsaari	8·50
Vaasa	7·60
Kaskinen	7·03
Mantyluoto	6·20
Rauma	4·93
Turku	3·53
Degerby	4·20
Hanko	2·73
Helsinki	1·83
Hamina	1·80

(Determined by author from data in Lisitzin, 1964).

changes. In essence the post-glacial eustatic changes have tended to lead to marine transgressions, whilst isostatic rebound from the weight of the ice caps has tended to lead to regression. In north-western Europe one can see how, through time, the rates of these two contrary processes has led to distinctive patterns.

In Scotland the rate and degree of isostatic rise was relatively small compared with parts of Scandinavia and Canada (see p. 187), so that at times, when the eustatic sea-level rise was going ahead with great speed (e.g. 14 000 to 6 000 B.P.) it overtook the isostatic rebound of the land and so resulted in a temporary transgression of the sea interrupting the emergence of the coast. Thus in certain parts of Scotland, notably the Firth–Clyde lowlands, there are alternating deposits of marine and freshwater sediments that comprise a valuable record of land and sea-level changes, and detailed survey, radiocarbon dating, and pollen-analysis allow a comprehensive picture to be presented (Donner, 1970, Walton, 1966).

Because of the weight of the ice the Late Weichselian situation was one of relatively high sea-levels and submergence. It was during this phase that many of the high level platforms were cut by the sea. The highest altitude of the Late Weichselian submergence is shown by raised beaches at 30–35 m above Ordnance Datum in Central Scotland. The Flandrian eustatic transgression, however, reached its maximum between 8000 and 6000 B.P. after a low position of sea-level between 10 000 and 8000 B.P. (see Fig. 6.10). During that regressive phase of low sea-level some distinctive peats were formed and these now often underlie marine deposits of the Flandrian transgression, such as the Carse Clay, which produced another suite of raised beaches at a level generally lower than those formed during the

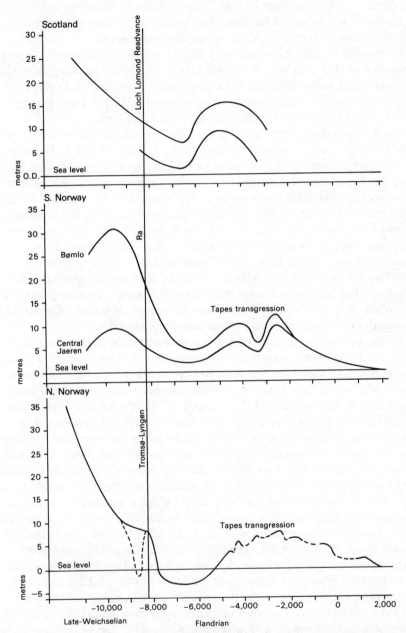

FIG. 6.10. Land/Sea-level changes in Scotland (A) compared with south Norway (B) and north Norway (C). The main moraines in each area are also included in the Figure (after Donner, 1970, Fig. 7).

maximum of the Late-Weichselian. This post-glacial shoreline reaches an altitude of 15 m in central Scotland, but because of the varied nature of the isostatic effect it declines in all directions from this peak.

In Norway the pattern of sea-level change in post-glacial times is essentially similar (Fig. 6.9) with high Late Glacial shorelines resulting from the ice-load effect being followed by a low stand of sea-level between about 12 000 and 10 000 B.P., and a transgression, often called the *Tapes* transgression in Norway and the *Nucella* transgression in Iceland.

Another situation where the effects of the post-glacial isostatic and eustatic sea-level changes can be well seen is in the development of the Baltic Sea.

At the maximum of the Weichselian Glaciation, 18 000 to 20 000 years ago, the Baltic area was covered by a great ice-body which deposited the Brandenburg and later moraines over the North European Plain. As this great ice sheet spasmodically retreated a series of small ice-dammed lakes developed in the southern part of what is now the Baltic. The coalescence and expansion of these lakes produced the first major stage of the post-glacial evolution of the Baltic—the Baltic Ice Lake, dated at about 11 000 B.P. This lake was not totally landlocked and managed to overflow to the low Early Holocene ocean level by a series of channels, including one into the White Sea. The location of the overflow channels varied according to deglaciation, glacial re-advances, and isostatic updoming.

In due course the deglaciation of southern Sweden enabled the sea to open the Baltic basin, and thus the Yoldia Sea was established at about 10 300 B.P., and lasted for less than 1000 years. However, the link between the Yoldia Sea and the ocean via southern Sweden was severed as isostatic updoming in that area raised the connection above sea-level. This thus led to the formation once again, after 9800 B.P., of a lake—the Ancylus Lake— which lasted for more than 2000 years, and its outlet was eventually through the Oresund, the channel that still separates Sweden from Denmark. Finally, the general world-wide rise in sea-level exceeded the rate of isostatic uplift and the Oresund was submerged, thereby changing the Ancylus Lake into the so-called Littorina Sea from about 7000 B.P. onwards. This sea was characterized by a relatively warm-water fauna, of a salt-water type, and was probably contemporaneous with the Tapes Sea recognized by Scandinavian workers outside the Baltic Basin. After around 3000 years (at about 4000 B.P.) the Littorina Sea merged into the current Baltic Sea, though faunal evidence suggests that as a result of the reduction in its outlet by updoming, the waters of the Baltic became progressively more brackish, with the associated constriction of the oceanic connection. Given present rates of uplift, however, and assuming no general rise of sea-level, the Baltic, currently linked to the ocean by channels only 7–11 m deep, could become a lake again in 8000–10 000 years.

Likewise the Black Sea was until around 9300 B.P. an oxygenated fresh-water lake separated from the Mediterranean by a sill in the vicinity of the present Bosphorous, converting the Black Sea before being converted into the anoxic marine habitat of today (Degens and Hecky, 1974). A combination of eustatic rise and subsidence in the southern North Sea led to the development of the North Sea as we know it now. Figure 6.11 illustrates how around 9500 B.P. the North Sea was largely dry land, and that the only sizeable body of water was the course to the south-west of the Rhine–Meuse system. However, by 8300 B.P. the situation had changed very radically, and the North Sea was then linked through to the English Channel. Compared

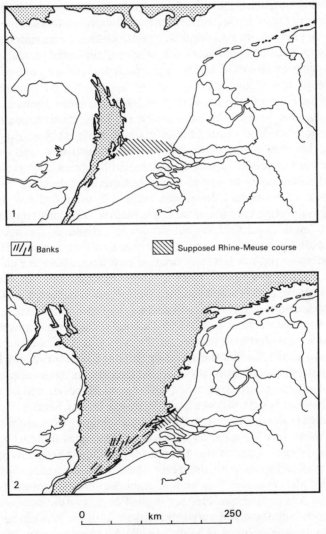

FIG. 6.11. The extension of the southern North Sea during the Early Holocene around 9300 B.P. (1) and 8300 B.P. (2) (after de Jong, 1967, Figs. 25 and 26).

to the present shore lines there remained extensive areas of low-lying ground at the mouths of the major estuaries, but these too were flooded as the Holocene progressed. The gradual sequence of inundation can be seen from an examination of the history of the European coastal lowlands.

Holocene sea-level movements and European lowlands

The various alternations of sea-level and climate in the Holocene had very marked effects, notably on the Dutch coast, the Fens of East Anglia, and the Somerset Levels. The sediments of these areas show the alternations that took place between marine, brackish, and freshwater sedimentation according to the degree of sea-level rise, sediment delivery by streams, and climatic conditions. These alternations had important consequences for human settlement, whilst the present-day arrangements of clays, silts, and sands, resulting from the complex Holocene history are reflected in current land utilization.

The complexity of the situation can be appreciated by a study of the history of the Fens and the Somerset Levels (Willis, 1961). In the case of the Fens the Jurassic floor to the area, smothered in places with boulder clay, was still above sea-level at about 5500 B.P. and drainage was largely good. The Fens then had a vegetation cover of large oak forest, and one fossil trunk from this period had a length of 20 m without branching. However, as the sea-level continued to rise in neolithic times and drainage conditions deteriorated, peat was laid down in a sedge fen, though this in turn was overlain by a marine clay (deposited in a shallow, brackish lagoon) which was about 7 m thick near the sea but steadily less inland. Radiocarbon dates from the peat beneath the clay suggest an age of about 4700 B.P. However, the transgression did not last long because peat from above the clay (often called the Buttery clay) gives dates of about 4200 B.P. inland, and about 4000 B.P. near the sea. A second marine transgression occurred in Romano-British times, and the peat (the top of which has been dated at A.D. 85 and 110 B.C.) was then overlain by silts. This second transgression appears to be synchronous with that encountered in the Somerset Levels (q.v.) and the Netherlands. Unlike the first transgression this one did not produce a large inland lake to be filled with brackish water but gave rise to extensive coastal salt marshes of silt and raised river banks or levées which penetrated far into the country. Such raised banks can still be traced in the Fenland landscape and are called *roddons*, and even in Roman times, judging from archaeological evidence, these were favoured occupation and road sites standing up above the level of the peaty countryside. Faunal analyses and the remains of whales and similar large marine animals suggest these rivers were indeed estuarine.

Work by the Fenland Research Committee has suggested that the Bronze-Iron Age marine transgression in the Fens may have been one of the prime reasons why there was very little settlement of the silts of the Fens in pre-Roman Iron Age times (Phillips, 1970). After about A.D. 80 a slight regression of the sea seems to have led to drier conditions in the area, and former estuarine flats emerged to a sufficient degree to be generally above the high-tide level. Tidal water would have been relatively confined in channels and the levées along such channels would have become suitable places for later Roman settlement and colonization. At this time the recently exposed land would have been largely free of forest cover and there would be no problems posed by long-established native ownership. Moreover, the Romans were not without experience of the immense assets of such coastal territory, reminiscent in certain respects of the Po marshes and many other river deltas throughout their Empire. However, on the departure of the Romans a combination of silting up of the rivers and of a deterioration in the artificial drains led to a reversion to less favourable conditions.

In Somerset the sequence is not dissimilar. At the base, one has the valleys formed during Pleistocene low sea-levels. Borehole records show that these valleys are filled with a blue-green clay to the approximate level of the present sea, and on faunal grounds it seems likely that this clay was deposited under brackish water conditions associated with a rising sea-level. This first transgression, which occurred earlier than in the Fens (from about 5500 B.P.) was succeeded by a period of freshwater marsh development during which *Sphagnum* bogs were developed on the waterlogged surfaces of the underlying marine clay. The period of bog growth, however, was not without its fluctuations, for in the Bronze Age (possibly 3000 to 2500 B.P.) an increase in wetness necessitated the construction of many wooden trackways (corduroy roads), since buried, which crossed from one upstanding island (such as the Isle of Avalon or Brent Knoll) to the next. It was, however, as in the Fens, inundated in Romano-British times (about 2000 yrs. B.P.), and more clay was deposited in the valleys. To what extent this marine inundation was due to an actual sea-level rise, and to what extent it might have been due to such factors as occasional high tides or storm surges, is a matter of some controversy, and has been discussed by Kidson (1977).

The Netherlands also show a similar sequence of regressions and transgressions, and the location and characteristics of the various sediments thereby produced are of considerable importance in terms of the present landscape and landuse of the country (de Jong, 1967). It has to be remembered that in discussing the Netherlands one is dealing with a more complex situation than existed in the relatively limited areas of the Fens and the Somerset Levels, but certain similarities are nevertheless evident for much of the country (Fig. 6.12).

The rising, post-glacial sea-level, accompanied by rises in temperature and of groundwater levels, produced environmental conditions favourable for a thriving vegetation, so that peat colonized the Pleistocene sandy and clayey deposits in Boreal and early Atlantic times. This peat accumulation started nearest to the coast first, and the Lower Peat horizon of this stage is some decimetres thick and occurs at about 12–20 m below sea-level outside

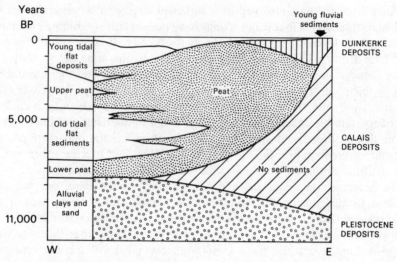

Fig. 6.12. The sequence of Holocene sediments in the Netherlands.

the flood plains of rivers in the coastal provinces. During Atlantic and early Sub-Boreal time parts of these provinces were affected by the rising sea-level which was relentlessly approaching that of the present. The peat layers of the Lower Peat were covered by marine and brackish water sediments, except in inland areas where the transgression was ineffective, and peat formation was able to continue uninterrupted so long as other environmental conditions were favourable. The marine and lagoonal sediments of the Atlantic and early Sub-Boreal transgression are called the Calais deposits, and consist of a humic or gyttja-like layer called the Velsen Clay and some sandier shoal-like deposits called the Old Tidal Flat deposits.

In late Atlantic and early Sub-Boreal times the rate of sea-level rise declined, and fluctuations of the relative power of marine and fluvial processes caused local regressions during which peat layers accumulated and some coastal barriers became established. In due course the formation of coastal barriers quickened and accretion behind them led to the drying up of the environment behind the barriers, a reduction in salinity, and establishment once again of terrestrial peat deposits, the Upper Peat horizon. A *Phragmites* peat tended to accumulate in the more brackish water, *Carex* peat occurred in areas of seepage, a eutrophic peat developed in areas bordering rivers, whilst an oligotrophic peat of *Sphagnum* occurred in places where peat growth depended mainly on rainwater.

The next phase in the evolution of the Netherlands coastal area was in Sub-Atlantic times, when the Upper Peat was replaced by the marine Duinkerke deposits, some of which are dated at 300 B.C. onwards (Duinkerke I), whilst others are dated at the late third century A.D. and at the ninth century onwards (Duinkerke II and III). The transgressions of early and late Roman times may be equivalent, as already stated, to those in Somerset and the Fens.

Reading for Chapter Six

The immense literature on sea-level changes has fortunately been put into an annotated bibliography, H. G. Richards and R. W. Fairbridge *et al.* (1970), *Annotated bibliography of Quaternary shorelines* (Supplement 1965–9), Special Publication No. 10, Academy of Natural Sciences, Philadelphia. The two best single reviews are A. Guilcher (1969), 'Pleistocene and Holocene sea-level changes', *Earth Science Reviews* 5, 69–98, and S. Jelgersma (1966) 'Sea-level changes in the last 10 000 years', in *Royal Meteorological Society International Symposium on World Climate from 8000–0 B.C.* A more controversial review with much good material is that by R. W. Fairbridge (1961), 'Eustatic changes in sea-level', *Physics and Chemistry of the Earth* 4, 99–185. Another publication which gives a large amount of data on the nature of sea-level changes in different areas is edited by A. Guilcher (1970), 'Symposium on the evolution of shorelines and continental shelves in their mutual relations during the Quaternary', *Quaternaria* 12 (entire volume). A valuable compilation of available dates on pre-Würm sea-levels is that by C. Lalou *et al.* (1971), 'Données géochronologiques actuelles sur les niveaux des mers et la paléoclimatologie de l'interglaciare Riss-Würm', *Revue de Géographie Physique et Géologie Dynamique* 13(5), 447–61. A penetrating review of the problems of dating and interpretation of interstadial and interglacial shorelines is that by B. G. Thom (1973), 'The dilemma of high interstadial sea-levels during the Last Glaciation', in *Progress in Geography* 5, 167–246. The information available specifically on post-Würm levels is greater, but the following are major contributions of more than local interest: N. A. Mörner (1969), 'The late Quaternary history of the Kattegatt Sea and the Swedish west Coast', *Sveriges Geologiska Undersökning Series C. NR.640, Arsbok 63, NR.3,* H. Godwin *et al.* (1958), 'Radiocarbon dating of eustatic rise in ocean level', *Nature* 181, 1518–19, J. D. Milliman and K. O. Emery (1968), 'Sea-levels during the past 35 000 years', *Science* 162, 1121–3, and N. C. Flemming (1969), 'Archaeological evidence for eustatic change of sea-level and earth movements in the western Mediterranean during the last 2000 years', *Geological Society of America, Special Paper 109.*

With regard to the causes of the observed changes of sea-level a simple treatment is that of C. G. Higgins (1965), 'Causes of relative sea-level changes', *American Scientist* 53, 464–76, but for a detailed discussion of the nature and effects of isostasy it is necessary to go to J. T. Andrews (1970), 'A geomorphological study of postglacial uplift with particular reference to Arctic Canada', *Institute of British Geographers Special Publication No. 2.* The relatively new appreciation of the importance of hydroisostasy is reviewed in A. L. Bloom (1967), 'Pleistocene shorelines: a new test of isostasy', *Bulletin Geological Society of America* 78, 1477–94, and (1969), 'Isostatic effects of sea-level changes' by C. G. Higgins in *Quaternary Geology and Climate,* ed. by H. E. Wright, 141–5. Orogeny as a possible cause of world sea-level change is proposed by R. L. Grasty (1967), 'Orogeny, a cause of world-wide regression of the seas', *Nature* 216, 779. The effects of various other tectonic and epeirogenic movements is being re-evaluated at the present time, and various symposia have been held of which the results have been published: 'Symposium on recent crustal movement', *Canadian Journal of Earth Sciences* 7 (2), 553–724, 'Recent crustal movements', ed. by B. W. Collins and R. Fraser (1971), *Bulletin Royal Society of New Zealand* No. 9, and 'Subsidence in south-east England' (1972), *Philosophical Transactions of the Royal Society, London,* A, 272. Other useful data on the same theme have been summarized in S. A. Schumm (1963), 'The disparity between present rates of denudation and orogeny', *United States Geological Survey Professional Paper 454–H.*

7 The Causes of Climatic Change

> It is not a field in which many people can dwell comfortably for a long time because it is almost entirely speculative. ... There is probably no 'right' theory to explain the cause of the Ice Ages, only a number of more or less probable ones.
>
> SPARKS AND WEST (1972, p. 26)

Introduction

The climatic changes that have been established and described, and which formed the basis for associated environmental changes such as those of sea-level described in the last chapter, have caused a great deal of discussion with regard to their causes. The purpose of this chapter is to summarize some of the main opinions that have been put forward, to stress the variety of factors involved, and to show the doubts still associated with the major hypotheses.

An indication of the complexity of factors that needs to be considered in any attempt to explain climatic change is given in Fig. 7.1. This flow diagram starts with the ways in which the input of solar radiation into the earth's atmosphere can fluctuate. For reasons such as varying tidal pull being exerted on the sun by the planets, the quality and quantity of outputs of solar radiation may change. The receipt of such radiation in the Earth's atmosphere will be affected by the position and configuration of the Earth and by such factors as the presence or absence of interstellar dust. Once the incoming radiation reaches the atmosphere its passage to the surface of the Earth is controlled by the gases, moisture, and particulate matter that are present. These materials may either be of natural or man-made origin. At the Earth's surface the incoming radiation may be absorbed or reflected according to the nature of the surface (the albedo). The effect of the received radiation on climate also depends on the distribution and altitude of land masses and oceans. These too are subject to change in a wide variety of ways—continents may move to or from areas where ice caps might accumulate, mountain belts may grow or subside to affect world wind-belts and local climates, and the arrangement of the climatically highly important ocean currents may be controlled by changes in sill depths and the widths of the seas, oceans, and channels. The situation is complicated, as the flow diagram suggests, by the existence of various feedback loops within and between the ocean, atmosphere and land systems.

In addition it needs to be remembered that the potential causative factors in climatic change operate over a very wide range of different time-scales, so

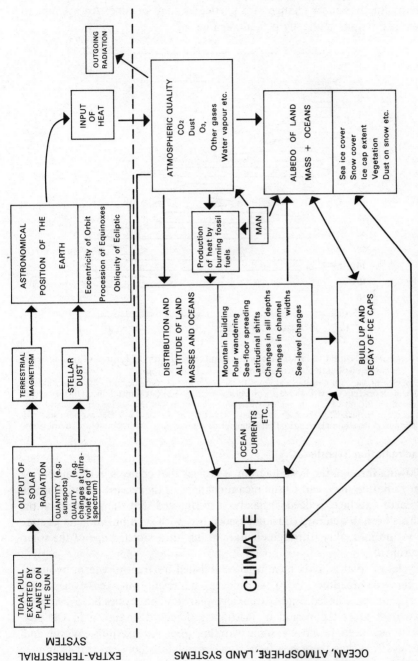

Fig. 7.1. A schematic representation of some of the possible influences causing climatic change.

that some factors will be more appropriate than others to account for a climatic fluctuation or change of a particular span of time. An attempt to show this diagrammatically is made in Fig. 7.2.

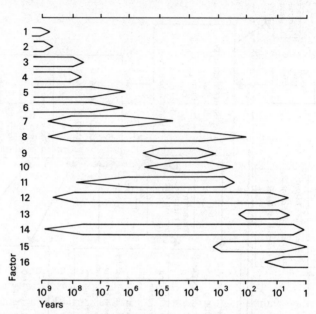

FIG. 7.2. Potential causative factors in climatic change and the probable range of time-scales of change attributable to each (after Mitchell, 1968, p. 156).

(1) Evolution of sun. (2) Gravitational waves in the Universe. (3) Galactic dust. (4) Mass and composition of air (except CO_2, H_2O and O_3). (5) Polar wandering. (6) Continental drift. (7) Orogeny and continental uplift. (8) CO_2 in the air. (9) Earth-orbital element. (10) Air-sea-ice cap feedback. (11) Abyssal ocean circulation. (12) Solar variability. (13) CO_2 in the air (fossil fuel combustion). (14) Volcanic dust in the atratosphere. (15) Ocean-atmosphere autovariation. (16) Atmosphere autovariation.

Solar radiation hypotheses

Following through the flow diagram, it is clear that changes in the output of solar radiation may lead to significant changes in the receipt of radiation at the Earth's surface. Indeed, it has been recognized that the sun's radiation changes both in quantity (through association with such phenomena as sun-spots) and in quality (through changes in the ultra-violet range of the solar spectrum).

Cycles of solar activity have been established for the short term by many workers (see Meadows, 1975), with eleven- and twenty-two-year cycles being ones particularly noted. Eighty to ninety-year sun-spot cycles have also been postulated. Over the period of instrumental record discussed in Chapter Five it has been found by some workers (see, for example, Wood and Lovett, 1974) that there has been some correlation between, for example, sun-spot activity and East African rainfall and lake levels. Sometimes, however, the correlations may suddenly break down, while other correlations may not necessarily be statistically significant. Nevertheless, some of the more significant associations may have predictive value. Thus, for example, Strongfellow (1974) plotted the five-year moving mean of lightning incidence in Britain against mean annual sun-spot numbers for 1930–73 and found a 0·8 correlation. He identified an eleven-year cyclic variation, with a trough for lightning being found in 1973. In that lightning is one of the main natural causes of electric power transmission failures in the United Kingdom such a relationship may assist the electricity authorities in planning maintenance services. At a less serious level a good correlation has now been found between sun-spot activities and achievement in sport. King (1973, p. 445) found that:

Information contained in Wisden can be used to show that, of the twenty-eight occasions on which cricketers have scored 3000 runs in a season in England, sixteen have been in sunspot maximum and minimum years; the five years in which this rare phenomenon occurred more than once were all sunspot minimum or maximum years. Likewise, thirteen of the fifteen occasions in which a batsman has scored 13 or more centuries in a season took place in, or within a year of, a sunspot maximum or minimum year.

Exceptional cricketing feats are produced at times of exceptional weather occurring at the extremes of the sun-spot cycle.

Although the role of changes in solar activity has frequently been attacked, especially with regard to cycles, some striking correlations have been found between changes in solar activity and certain major characteristics of the general atmospheric circulation. Figure 7.3, for example, shows a distinct similarity in trend between Baur's solar index and the yearly frequency of the general westerly zonal type over the northern hemisphere, with a

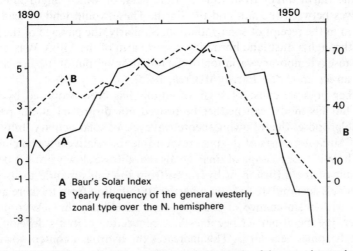

FIG. 7.3. Curves of Baur's solar index and the yearly frequently of the general westerly weather type over the northern hemisphere (after Lamb, 1969).

general increase in both parameters until the 1930s from 1890, and with a rapid decrease in both into the 1960s. This suggests that a portion of the observed climatic variation of the twentieth century may be attributable to a variation of the sun's energy output at source. Claims may also be made, however, for the importance of other factors.

Over a longer time-scale it is far more difficult to suggest that the sun's output of radiation has changed sufficiently to affect the Earth's climate, for any substantial proof is lacking. Nevertheless, this is a possible hypothesis, and one which has received much support. Some evidence to support it comes from studies of the oscillation in the concentration of atmospheric C14, which in turn depends partly upon variation in the emission of solar radiation. C14 levels have fluctuated during the Holocene, and Denton and Karlen (1973) have argued that the major intervals of high atmospheric C14 activity coincide with periods of neo-glacial expansion, while the intervals of relatively low C14 activity coincide with intervals of glacier contraction. Equally, Bray (1970) has suggested that Holocene glaciations show a periodicity of around 2600 years, and that an arithmetic progression starting with 22 years (the complete sunspot or 'Hale' cycle), and a first term of 4, results in a sequence of 88 440 and 2640 years. Other workers, utilizing spectral analysis of an ice core from Camp Century, Greenland, have claimed to find systematic long-term oscillations of a broadly comparable magnitude to those of Bray: 78, 181, 400, and 2400 years. These they also relate to varying solar activity (Johnsen et al., 1970).

The causes of variations in solar activity are still imperfectly understood, but one possible cause of variability in receipt of insolation on the Earth's surface is the presence of clouds of fine interstellar matter (nebulae) through which the Earth might from time to time pass, or which might interpose themselves between the sun and the Earth. These would tend to lead to a reduction in the receipt of solar radiation. Similarly, the passage of the solar system through a dust lane bordering a spiral arm of the Milky Way galaxy might cause a temporary variation of the radiation output of the sun, and so lead to an ice epoch on Earth (McCrea, 1975).

Another possible cause of solar variations has been proposed by Opik (1958), but his model can neither be proved nor disproved at the present time. He proposes the following theoretical cycle of solar activity. Initially a 'normal' situation exists of the type responsible for relatively warm climates on Earth. With the passage of time metals that diffuse slowly are left behind as a result of the diffusion of hydrogen from the sun's mantle to its core. These metals accumulate to form a barrier to radiation from the core and in keeping with maintenance of a steady-state condition, the sun contracts. However, the metal barrier becomes hot, convection currents develop, and the core becomes very large. This increases the hydrogen content available for fuel and the energy output increases. The production of heat is such that it cannot be adequately transported to the surface: thus the sun expands. In expanding, energy is expended thereby reducing the heat and light output from the sun. This gives reduced radiation and cooling on Earth. However, expansion lowers the temperature of the core and the amount of energy it produces. Thus the core shrinks and eventually the sun returns to its 'normal' position, giving relative warmth on Earth.

Climatic change and variations in terrestrial magnetism

In recent years a large quantity of work has been initiated into the relationship between changes in the intensity of the Earth's magnetic field and changes in climate. The work is in its early stages, but already some strong relationships have been established between temperatures, over time-scales of ten years to 1·2 million years, and magnetic intensity. For example, Wollin and his co-workers (1971, 1973) have found that over the period 1925–70 magnetic intensity has been decreasing at observatories in Mexico, Canada, and the United States at the same time as temperatures have been increasing. Equally, at observatories in Greenland, Scotland, Sweden, and Egypt the intensity is increasing whereas the climate is getting colder. In other words there appears to be a close inverse correlation between changes of the Earth's magnetic field and climate (Fig. 7.4).

Why this situation should exist is not clear. It is possible that the Earth's magnetic field changes in response to changes in solar activity and that both climate and terrestrial magnetism are yoked together in their response to solar events (Wollin et al., 1974). If this is the case then magnetism does not in itself have a simple cause and effect relationship with climate. On the other hand, it is possible that magnetism may modulate climate to some degree by the ability of the Earth's magnetic field somehow to provide a shield against solar corpuscular radiation.

Thus the relationship of these two phenomena appears to be proven, but the reason for the relationship is still not clear.

Earth geometry theories—the Croll–Milankovitch hypothesis

Following through Fig. 7.1, it is reasonable to assume that if the position and configuration of the Earth as a planet in relation to the sun were to change, so might the receipt of insolation from the sun. Such changes do take place, and there are three main astronomical factors which have been identified as of probable importance, all three occurring in a cyclic manner (Fig. 7.5): changes in the eccentricity of the Earth's orbit (a 96 000 year cycle), the precession of the equinoxes (with a periodicity of 21 000 years), and changes in the obliquity of the ecliptic (the angle between the plane of the Earth's orbit and the plane of its rotational equator). This last has a periodicity of about 40 000 years.

The Earth's orbit around the sun is not a perfect circle but an ellipse. If the orbit were a perfect circle then the summer and winter parts of the year would be equal in their length. With greater eccentricity there will be a

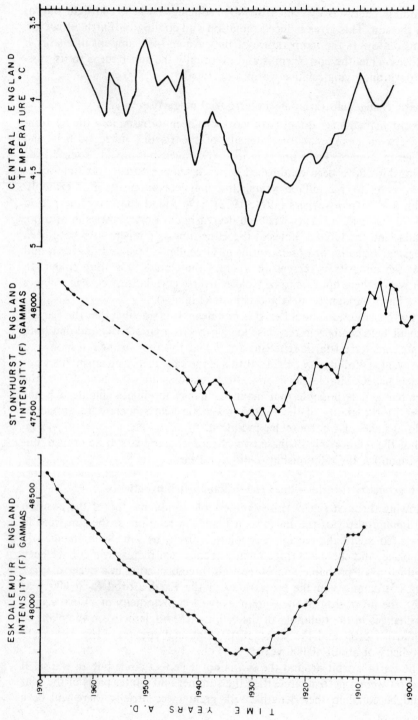

FIG. 7.4. Magnetic intensity curves based on annual means compared with 10-year running averages of winter temperature for central England (1900–70) (from Wollin et al., 1974, Fig. 6).

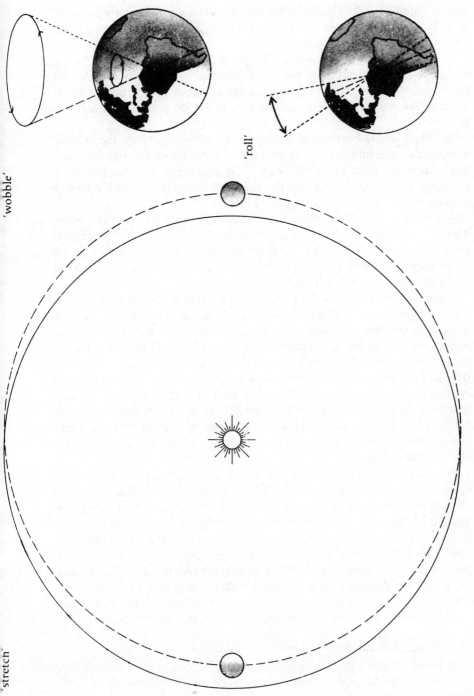

'wobble'

'roll'

'stretch'

Fig. 7.5. The three types of fluctuation in earth geometry involved in the Croll–Milankovitch hypothesis (after Calder, 1974).

greater difference in the length of the seasons. Over a period of about 96 000 years the Earth's orbit can 'stretch' by departing much further from a circle, and then revert to almost true circularity.

The precession of the equinoxes simply means that the time of year at which the Earth is nearest the sun (perihelion) varies. The reason is that the Earth wobbles like a top and swivvles its axis round and round. At the moment the perihelion comes in January. In 10 500 years it will occur in July.

The third cyclic perturbation, change in the obliquity of the ecliptic, involves the variability of the tilt of the axis about which the Earth rotates. The values vary from 21°39′ to 24°36′. This movement has been likened to the roll of a ship. The greater the tilt, the more pronounced is the difference between winter and summer (Calder, 1974).

Appreciation of the possible significance of these three astronomical fluctuations of the Earth goes back to at least 1842, when J. F. Adhémar made the suggestion that climate might be affected by them. However, his views were developed by Croll in the 1860s and by Milankovitch in the 1920s (Beckinsale, 1965; Mitchell, 1965).

The major attraction of these ideas is that while the amount of temperature change caused by them may well only be of the order of 1 or 2 °C, the periodicity of these fluctuations seems to be largely comparable with the periodicity of the ice advances and retreats of the Pleistocene. Recent isotopic dating has shown that the record of sea-level changes preserved in the coral terraces of Barbados and elsewhere, and the record of heating and cooling in deep-sea cores, correlates well with theoretical insolation curves based on those of Milankovitch (Messollela *et al.*, 1969; Broecker *et al.*, 1968). Thus there is some substantial evidence to suggest that the astronomical theories may be valid as an explanation of the longer scale of environmental changes.

However, the Croll–Milankovitch model produces a cyclical series of events which is too long to be relevant to post-glacial fluctuations of climate, and too short to throw much light on the spacing of the major ice ages. In addition, the model advocates the development of glaciation in high latitudes by insolation variations, whereas from a glacial mass budget viewpoint, an increase in precipitation over the present minimal levels now received in polar areas may be more important (Andrews, 1975). Finally, the computed variations in insolation resulting from this model are never more than a very few per cent so that it is likely that even if this mechanism can initiate change, other factors would be necessary to intensify it.

Atmospheric transparency hypotheses

Even if one supposes that changes in receipt of insolation from the sun have not taken place to a sufficient degree markedly to change the Earth's climate, the effects of incoming radiation from the sun may have been altered markedly by changes in the atmospheric composition of the Earth. This could take place through changes in the carbon dioxide levels, ozone levels, dust levels, and water content.

Consideration of the possible role of carbon dioxide is made with regard to the Plass Hypothesis (see p. 214) and with regard to the role of man as an agent of climatic change in recent years (see below p. 215). It needs to be pointed out, however, that the likelihood of very substantial changes in the carbon dioxide content of the atmosphere is subject to some doubt because the Earth's carbon cycle is controlled largely by oceanic absorption of the gas. The oceans form a vast reservoir of carbon compounds. Moreover, at present it is difficult to see what factors could have caused a sufficient degree of change in CO_2 contents in the past. Nevertheless, other things being equal, an increase in the carbon dioxide content of the atmosphere, because this would lead to an increased absorption of long-wave terrestrial radiation in the thirteen to seventeen micron band, would tend to promote an increase in world temperatures.

Ozone in the upper stratosphere at 30 to 50 km is effective in the screening of incoming solar radiation (by short-wave absorption) and may affect outgoing or terrestrial radiation by infra-red absorption. Changes in ozone concentrations, however, may themselves be caused by variations in solar emission. In general increases would tend to lead to increases in surface temperature.

Volcanic eruptions could lead to a cooling of climate by the production of a dust-veil in the lower stratosphere (Lamb, 1970). The time-scale involved would probably be short so that their significance may be limited to secondary and minor fluctuations of climate. However, recent studies of eruptions and temperature do suggest that over a very short time-scale they may be important. The ash emissions of Krakatoa in the 1880s and Katmai (1912) produced a global decrease of solar radiation of 10 to 20 per cent lasting 1 to 2 years, and the ash from Krakatoa was injected into the stratosphere, reaching a height of 32 km. It has recently been shown that many of the coldest and wettest summers in Britain, such as 1695, 1725, the 1760s, 1816, the 1840s, 1879, 1903, and 1912 occurred in conjunction with times of high volcanic dust inputs into the stratosphere and upper atmosphere (Lamb, 1971). Moreover, the period of temperature warming in the northern hemisphere that has been experienced in the 1920s, 1930s, and 1940s coincides with a period when there were no volcanic eruptions in the northern hemisphere, suggesting the possibility that the absence of a volcanic dust-pall in those decades was one factor in the warming process. Going back further, a study of the Byrd ice core from Antarctica has produced evidence of particularly heavy and frequent volcanic dust falls at 20 000 to 16 000 yrs. B.P.—the same time as the maximum cold of the Last Glacial (Gow & Williamson, 1971). Similarly, the little climatic optimum and the Little Ice Age (Bray, 1974) seem to correspond to periods of low and high volcanic activity respectively (Fig. 7.6).

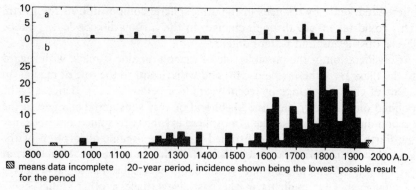

FIG. 7.6. Great volcanic eruptions in Iceland and incidence of Arctic sea-ice in that area since
A.D. 870.
A = great volcanic eruptions
B = polar ice at the coast of Iceland.
weeks/year, 20-year averages
(After Lamb, 1972, Fig. 10.22.)

In addition to the role of backscattering already referred to, volcanic dust
could reduce sunshine totals further by promoting cloudiness, for dust particles, by acting as nuclei, can promote the formation of ice crystals in
subfreezing air saturated with water vapour.

During the Holocene as a whole Bray (1974) has suggested on the basis of
examination of C14 dates that the major advances of alpine and polar
glaciers were exactly contemporaneous with the major post-Wisconsin volcanic activity phases in New Zealand, Japan, and southern South America
(4700–5450 B.P., 2150–2850 B.P., and 50–470 B.P.). Additional support for
the role of volcanism comes from the analysis of coarse volcanic ash in 320
deep-sea cores. Kennett and Thunell (1975) find that such ash is very
frequent in the Quaternary, being about four times the neogene average.

Hypotheses involving changes in terrestrial geography
Although some climatic changes take place over a short time-span, such as
the Little Ice Age or the twentieth-century phase of warming up, some of
the longer-term changes, including possibly the initiation of glaciation in
certain parts of the world, may have resulted from changes in the positions
of the continents, shifts in the position of the polar axis, and uplift of
continents by orogeny. Of these the first two are probably relatively unimportant in terms of the Pleistocene, in that the rates of change involved are
very slow and over the time-span concerned rather slight. The rate of polar
wandering, for example, has been estimated as 3×10^{-7} degrees yr.$^{-1}$, and
this would be insufficient to affect the pattern of Pleistocene glaciation
(Cox, 1968). The rate of continental drift is a little higher, with a mean rate
of around $1 \cdot 0 \times 10^{-7}$ degrees yr.$^{-1}$, which is equal to a displacement of
only 1° during the last 10^7 years (possibly only $0 \cdot 2°$ since the start of the

classic glaciations). Even with a maximum postulated rate of $6 \cdot 0 \times 10^{-7}$ degrees yr.$^{-1}$, the displacement is not of very great significance. However, Ewing (1971) has recently suggested that if sea-floor spreading operated at a rate of about 2 cm/yr. the width of a rift such as that between Spitzbergen and Greenland would increase by about 200 km in 10 m.y., sufficient to affect the entry of ocean currents into the Arctic and thereby the climate of the surrounding areas. Nevertheless, many workers consider that terrestrial causes of climatic change can be narrowed down to vertical uplift of mountains through orogeny so that their summits become sufficiently high and cold for snow and ice to accumulate. This could well have important effects, for as pointed out on p. 188 there has been considerable tectonic movement in many areas in the Pleistocene and Late Tertiary.

If one assumes a rate of uplift in a tectonically highly active area of 10 m per 1000 years it would only require about 10 000 years to obtain a mean temperature fall of 0·65 °C, since temperature falls at a rate of about 0·65 °C with every 100 m rise in altitude. Thus in the course of the Pleistocene it would be quite possible for a mountain to develop sufficiently quickly to cause a marked temperature depression at its summit. Also, total precipitation is known to show a general increase with altitude, at least up to about 3000 m, so that the over-all increase in height of mountains would serve to produce veritable snow traps. However, if elevation alone were the dominant cause of glaciation a large snow or ice field thus established would tend, through its effect on albedo and on pressure systems, to be self perpetuating, and so some other mechanism must be introduced to account for the disappearance of the ice mass.

The effects that would result from such uplift would be both local and global, in that, for example, the uplift of the Rockies could well affect the whole weather situation in the northern hemisphere by its effect on standing atmospheric waves and blocking anticyclones across the north Atlantic. Some support for this hypothesis is given by the fact that not all areas appear to have undergone multiple glaciation, and in many cases it seems possible or probable that uplift in the mid and Late Pleistocene brought the mountains of some areas into a position where ice could accumulate. Thus, for example, Mauna Kea (Hawaii), Tasmania, and the Pyrenees have been cited as areas which anomalously only suffered one major glacial phase, and that late in the Pleistocene.

Feedback (autovariation) hypotheses

Thus far in our consideration of the possible causes of climatic change we have suggested that it has been caused by significant changes in the output of solar radiation, the position and configuration of the Earth in relation to other heavenly bodies, the quality of the atmosphere, and the arrangement of landmasses, oceans, and mountains. However, there have been a variety of hypotheses in which it is envisaged that the atmosphere possesses a degree of internal instability which might furnish a built-in mechanism of change. It is conceivable that some small change might, through positive feedback, have extensive and long-term effects. As Mitchell (1968, p. 158) put it, 'minor environmental disturbances may have sufficed to "flip" the atmospheric circulation and climate from one state to another, and to "flop" it back again'. Some selected examples will give an indication of the importance of the hypotheses involving feedback relationships.

Wilson's (1964) hypothesis

At a time when the total thickness of the ice cap of the Antarctic is less than a critical value, the rate of thickening produced by the accumulation of precipitation exceeds the rate of subsidence produced by plastic flow of the ice cap and by mass loss through calving at the perimeter. When, however, the ice thickness reaches a critical threshold value, the transverse shearing stress near the base of the ice cap becomes so large that the ice flow abruptly accelerates. This produces heat by friction, so that flow accelerates further until the whole ice cap subsides at a more or less catastrophic rate. This fills the world oceans with cold ice and thereby leads to a reduction in world temperatures, which promotes glaciation in certain other parts of the world (Hollin, 1965; Selby, 1973).

In addition, as a result of the surge of the ice cap, possibly one third of the ice sheet is transferred onto the continental shelf to form a huge ice shelf. This shelf would increase the surface albedo of 25×10^6 km^2 of ocean from 8 per cent to 80 per cent, adding to the cooling effect by decreasing the heat input to the Earth as a whole by 4 per cent.

The Plass Hypothesis (1956)

An unspecified cause reduces the carbon dioxide content of the atmosphere by several per cent. This would lower atmospheric temperatures, and after 50 000 years or so the oceans would cool by a similar amount and come to a new CO_2 equilibrium with the atmosphere. The lowered temperatures would promote continental glaciation, which would in turn lower sea-levels, and thus cause a new imbalance of carbon dioxide with the atmosphere by increasing its concentration in the oceans. The increased CO_2 content of the atmosphere leads to atmospheric warming, leading to glacier melting, and a restoration of the oceans to their original volume.

The Ewing-Donn Hypothesis (1956, 1958)

The cycle of events begins with high interglacial sea-levels, with a flow of warm water into the Arctic ocean, which both keeps that ocean ice free and favours the accumulation of precipitation in the form of snow on the surrounding land masses. This lowers the sea-levels so that a submarine ridge between Iceland and the Faeroes begins to block the further flow of warm water into the Arctic. The increased area of the growing ice cap would also lead to a greater reflection of solar radiation. This would accelerate the rate of cooling. Such a tendency would be further reinforced by the formation of an anticyclone over the ice, with outblowing winds repelling the moderating influence of Atlantic conditions. Thus the Arctic freezes over and prevents any on-going replenishment of the ice sheets which thus gradually waste away. Sea-level thus rises and warm water flows in again, starting the cycle off once more.

Recent work on Arctic ocean cores, however, suggests that the Arctic ocean was never ice free in the Pleistocene and so could not have been a factor in the growth or melting of Pleistocene continental glaciers (Larsen and Barry, 1974, p. 258).

Albedo-based hypotheses

One factor which controls heating levels in the earth-atmosphere system is the degree to which incoming solar radiation is reflected or absorbed by the Earth's surface. Changes in the albedo of that surface, brought about perhaps by seemingly minor events, might be able to cause major changes in climate. For example, the deposition of a layer of relatively dark volcanic dust over the ice caps, as a result of a chance volcanic explosion, might lead to the melting of that ice cap which might in turn set a train of events in motion. Similarly, the presence of an unusually widespread and persistent snow cover over northern Canada as a result of a chance association of snowy winters and cool summers, could either help to trigger off climatic change directly, or could play a role as part of a feedback reaction (Williams, 1975). Such a snow cover, persisting through all or most of the summer and autumn, would reflect the sun and further chill the air (Calder, 1974). This in itself might increase the likelihood of snow the next winter. The snow gradually accumulates leading to a great ice sheet.

Man's effect on climate

The miscellaneous hypotheses discussed so far have been applied with varying degrees of success to a variety of different time-spans. When, however, one considers the immediate past and contemplates the near future, the role of man takes on a position of probable importance For, although as explained in Chapter Five, the twentieth-century climatic changes have greatly affected man, it is more than likely that man has himself been partially responsible for some of the observed changes, particularly because of his effect on atmospheric quality. As yet, because of the complexity of the atmospheric system and the large number of possible causes, it is difficult fully to assess and quantify the role that man has played, though certain mechanisms of man-induced climatic changes on a global (as opposed to a microclimatic scale) can be recognized.

One of the most important mechanisms is man's consumption of fossil fuels including coal and oil. Until recently the amount of energy used by man has been negligible compared both to the resources of solar energy and to the energy of photosynthesis by plants. However, that situation has begun to change. World energy consumption is increasing at a rate of around 4 per cent per annum, equivalent to a doubling every seventeen years (Budyko et al., 1971).

Connected intimately with this production of heat, which on a local scale can be identified as the 'urban heat island', is the build up of CO_2 concentrations in the atmosphere. At present the rate of CO_2 build up is about 7 ppm per decade (Sawyer, 1971) and the 1960 CO_2 concentration was 313 ppm. The concentration of CO_2 affects the amount of solar radiation reaching the earth and in general an increase should lead to a warming tendency. It has been estimated that a doubling of CO_2 would raise surface temperatures by around 1·3 °C. However, some recent observations and investigations suggest that the rate of increase in temperature diminishes with increasing atmospheric CO_2 contents, so that temperature increases are unlikely to attain a high level.

Also following on from the consumption of fossil fuels is the increased emission of various pollutants into the atmosphere. An increase in dust or smoke particles would have an effect on the scattering and absorption of solar radiation so that world temperatures would tend to change. It might also cause a reduction in rainfall by causing a reduction in convective activity (Bryson and Barreis, 1967), though conversely it has been argued that an increase in particulate matter in the atmosphere might provide further nuclei for condensation and sublimation of water vapour in the atmosphere, thereby causing an increase in cloud cover (see Gribbin, 1975 for a discussion). Unfortunately the precise effects of aerosols on temperature are still not clear, for whether added aerosols cause heating or cooling of the earth-atmosphere system is a function not only of their intrinsic absorption-backscatter characteristics, but also of their particular location in the atmosphere with respect to cloud-cover, cloud reflectivity, and underlying surface reflectivity (Weare et al., 1974). Thus it may be that near the poles 'grey' aerosol particles would cause warming of the atmosphere because they would be less reflective than the underlying white-ice and snow surfaces, whereas over a dark vegetated surface they would reflect a relatively greater amount of radiation, leading to a cooling (Peck, 1975). Thus the quantitative effects of an increased atmospheric aerosol content are unclear, but Rasool and Schneider (1971) have suggested that an increase by only a factor of 4 or 5 in global aerosol concentrations could be sufficient to reduce surface temperatures by as much as 3·5 °C. Fortunately, the developed countries, the countries which produce the bulk of the atmospheric aerosols of non-natural origin, have the technological resources largely to overcome this problem. Indeed, some of them have already taken steps in this direction. Nevertheless, a striking illustration of increases in dust incidence since the beginning of the Industrial Revolution is provided by the analysis of dust levels in glacier ice of known age in the southern U.S.S.R. Layers of ice dated A.D. 1800–1920 show a dust content of around 10 mg/l whereas by the 1950s the figure had increased twentyfold to 200 mg/l (Davitaya, 1969).

Another celebrated issue related to the effects of man on atmospheric quality and thereby potentially on climate is the role of chemicals, particularly the chlorofluoromethanes, emitted when household aerosol cans are used. It has been suggested that their chemical inertness and high volatility (qualities responsible for their use) mean that they remain in the atmosphere for a long time and therefore build up to relatively high levels. It is thought that photodissociation of these gases in the stratosphere produces significant amounts of chlorine atoms, and leads to the destruction of some atmospheric ozone. Ozone, as already noted, is an important control of radiation.

Another potentially serious problem is that involving high-altitude aircraft and rockets (Report of the Study of Critical Environmental Problems, 1970, p. 17). The latter will tend to introduce exotic chemicals into the high atmosphere through their exhausts. It is known that even small quantities of an element like ozone in the upper layers of the atmosphere can markedly control radiation conditions. Thus small additions to this zone, or reactions involving the addition of exotic chemicals, could potentially have major results. However, in the short term the water vapour discharged by supersonic aircraft into the stratosphere may be more serious. At the present the water content of the stratosphere is low, and the exchange of air between the lower stratosphere and other regions is low. Consequently, comparatively modest amounts of water vapour discharged by aircraft could have a significant effect on the natural balance. It has been calculated that 400 supersonic aircraft, whether military or civil, making four flights per day would place 150×10^6 kg of water into the lower stratosphere (Sawyer, 1971). Over a period of years this could double the existing content of water in the stratosphere. Such an increase would lead to a small rise in temperature, perhaps 0·6 °C. The presence of this moisture might also be detectable in the form of thin high-altitude cirrus, or cirro-stratus.

On the continental or regional scale, it has often been maintained, especially in the pre-war years, that afforestation would improve the rainfall conditions, notably of desert margins, and that, conversely, deforestation would lead to a decline in rainfall conditions. Thus through his influence on forest cover in zones like the Sudan zone of West Africa, man was seen as a possible cause of some of the desertification that was alleged to have taken place (see p. 149).

Assertions of this nature were based on the established fact that the presence of a forest has a favourable influence on the water economy of an area. This phenomenon was first attributed to increased precipitation. Moreover, the high humidity within forests, the observation of 'forest smoke' over closestands, and the presence of high humidity in the air surrounding a forest, may all have provided apparent support for such propositions.

However, although schemes have been mooted to improve rainfall conditions on the Saharan fringes by means of afforestation of a great belt of land across West Africa, it has to be stressed that precipitation formation is an upper atmosphere process in many respects. So long as the major arid zones of the world are dominated by subsiding air, any slight humidity increases brought about by the presence of forest belts will be largely ineffectual. The same point probably holds for plans to establish large lake bodies in the Kalahari and the Sahara. The African coast of the Mediterranean provides a clear example of how little influence even a massive supply of warm vapour from a warm sea may have. The coasts remain arid because of the nature of the general circulation.

Nevertheless, although forests might not cause significant changes in precipitation through the mechanism of transpiration of moisture, there has been increased interest recently in the possible consequences of deforestation on climate through the effect of albedo change. Ground covered by plants has an albedo in the range of 10–25 per cent, whereas ground deprived of a vegetation cover as a result of deforestation and overgrazing (as in parts of the Sahel) has a very much higher one. This would affect temperature levels. ERTS satellite imagery of the Sinai-Negev region shows a very great difference in image between the relatively dark Negev and the very bright Sinai-Gaza Strip area. This line coincides with the 1948–1949 armistice line between Israel and Egypt, and results from land-use differences. Otterman (1974) has suggested that this land-use changed albedo has produced temperature changes of the order of 5 °C. However, the consequences may go beyond changing temperature. Charney and others (1975) have argued that the increase in surface albedo resulting from a decrease in plant cover, would lead to a decrease in the net incoming radiation, and an increase in the radiative cooling of the air. As a consequence, they maintain, the air would sink to maintain thermal equilibrium by adiabatic compression, and cumulus convection and its associated rainfall would be suppressed. The lower rainfall would in turn have an adverse effect on plants and lead to a further decrease in plant cover. Such considerations are plainly of importance if substantial deforestation occurs in the Amazon basin, and possible effects of albedo change in that area have recently been simulated in a computer model (Potter et al., 1975). However, this view is not universally accepted. Ripley (1976), for example, suggests that Charney and his co-workers, while considering the effect of vegetative changes on albedo, completely ignored the effect of vegetation on evapo-transpiration. He points out that vegetated surfaces are usually cooler than bare ground because much of the absorbed solar energy is used to evaporate water, and concludes from this that protection from overgrazing and deforestation, might, in contrast to Charney's views, be expected to lower surface temperatures and thereby reduce rather than increase convection and precipitation.

Conclusion

No completely acceptable explanation of climatic change has ever been presented, and it is also clear that no one process acting alone can explain all scales of climatic change. Some coincidence or combination of processes in time is probably required, such as Flint's 'solar-topographic' theory (1971), which depends mainly on variations in the intensity of solar radiation and of mountain building. Moreover, numerous feedback loops may exist. Some hypotheses appear plausible to explain variations over a long time period (for example the Croll-Milankovitch hypothesis may be applicable to glacial-interglacial cycles), while others appear more plausible for short-term fluctuations (changes in sun-spots may be a hypothesis relevant at a scale of a decade or more). Two other basic problems exist. One is that to test certain hypotheses we need precise knowledge of the exact pattern and dating of past fluctuations—this we seldom have. The second is that we are dealing with an immensely complex series of interrelated systems; the solar system, the atmosphere, the oceans, and the land. It is thus unlikely that any simple hypothesis or model of climatic change will have very wide applicability.

Given these considerations it is clearly impossible at the present state of knowledge to make any safe prognosis of the climatic developments of the future. Many such predictions have been made in recent years, but they seldom show much similarity. Calder (1974) and others have proposed that we are on the verge of a new ice age which will arrive with great suddenness; Winstanley (1973) and others have suggested that monsoonal areas will get progressively drier for some decades; while others have suggested that on account of man's activities temperatures are likely to rise very sharply, possibly to a level warmer than for a thousand years by the first decade of the next century (Broecker, 1975).

Some workers have attempted to predict on the basis of establishing the presence of cycles connected with solar activity or other phenomena, and a large number of cycles have been identified (see, for example, Table 7.1). It is salutary to remember, however, that such cycles have been mooted for a very long time: Sir Francis Bacon reported a thirty-five-year cycle of weather three and a half centuries ago. Ellsworth Huntington was probably right when he stated in *Mainsprings of Civilisation* (1945, p. 464):

It will be a vast boon to mankind when we learn to prophesy the precise dates when cycles of various kinds will reach definite stages.This would be easy if (1) there were only a few cycles; (2) each were absolutely uniform in length and intensity; (3) no cycle produced delayed effects or interfered with any other; and (4) a given cycle were equally developed in all parts of the earth. Not one of these conditions exists.

Table 7.1

Period lengths of some selected natural phenomena cycles

Phenomenon	Period lengths (years)
Ice core from Greenland	2400, 400, 181, 78
Post-glacial glaciations	2600
Lapland tree-rings	200, 90, 30, 23, 11–12
Crimean lake varves	11·2
Thunderstorms	11
Formosan tree-rings	100, 20–22, 11, 6
Droughts in the Great Plains	91, 46, 23
Salinity of Baltic, ice on Barents Sea, sea-level of Atlantic	80–90
Baltic Ice	21–24, 11–14, 8, 5–6, 3

Source: Lamb, 1972

Caution is desirable. As Mason (1976, p. 51) has averred in a recent review of the question of prediction of climatic change,

. . . warnings of an imminent ice age and other major catastrophes are ill-founded and irresponsible. The recent droughts in Africa, floods in Pakistan and tropical storms in Australia all have parallels in the past and do not imply that the global pattern of climate is undergoing a radical and permanent change. A more realistic if less dramatic assessment would be that fluctuations of climate will recur with about the same magnitude, frequency, and variability as in recent centuries, superimposed on longer-term trends whose onset and reversal cannot yet be accurately predicted.

An equally sober assessment has been made by Landsberg (1976, p. 442) in the course of an attempt to review two recent popular books, one of which proposed the imminence of great cold, the other the inevitability of immense heat: 'If you think you can extrapolate climate stick around for a while and learn better.'

Reading for Chapter Seven

There are three excellent surveys of theories on climatic change which have been produced in recent years. In *Essays in Geography for Austin Miller* (ed. J. B. Whittow and P. D. Wood, University of Reading Press), 1965, pp. 1–38, R. P. Beckinsale contributes 'Climatic change: a critique of modern theories'. Another review, of a similar type but with greater concentration on the auto-variation ideas is J. M. Mitchell (1965) 'Theoretical palaeoclimatology' in H. E. Wright and D. G. Frey (eds.), *The Quaternary of the U.S.A.*, pp. 881–901. J. Mitchell (1968) has also edited 'Causes of climatic change', *Meteorological Monographs* 8. These three sources give lengthy bibliographies and summaries of the major theories.

Since these works were published there has been a series of papers on the role of volcanic eruptions in causing climatic changes. The most comprehensive is by H. H. Lamb (1970), 'Volcanic dust in the atmosphere: with a chronology and assessment of its meteorological significance', *Philosophical Transactions Royal Society London* A, 266, 425–533. A shorter paper by the same author, 'Volcanic activity and climate' appears in *Palaeo* 10 (1971), 203–30, while an interesting short paper on the possible connection of the Antarctic volcanic eruptions with the Würm maximum is A. J. Gow and T. Williamson (1971), 'Volcanic Ash in the Antarctic ice sheet and its possible climatic implications', *Earth and Planetary Science Letters* 13, 210–18.

Similarly there has been a marked development of interest in the role of man as an agent of global climatic change. On the role of the increasing carbon dioxide concentrations see G. N. Plass (1959), 'Carbon dioxide and climate', *Scientific American* 201, 41–47; W. Bischof and B. Bolin (1966), 'Space and time variation of the CO_2 content of the troposphere and lower stratosphere', *Tellus* 18(2), 155–9; and S. I. Rasool and S. H. Schneider (1971), 'Atmospheric carbon dioxide and aerosols: effects of large increases on global climate', *Science* 173, 138–41. With regard to the role of aerosols, R. J. Charlson and M. J. Pilar (1969), 'Climate: the influence of aerosols', *Journal Applied Meteorology* 8, 1001–2, draw different conclusions to Rasool and Schneider. Other useful studies on aerosols include, F. F. Davitaya (1969), 'Atmospheric dust content as a factor affecting glaciation and climatic change', *Annals Association American Geographers* 59, 552–60; R. A. McCormack and J. H. Ludwig (1967), 'Climate modification by atmospheric aerosols', *Science* 156, 1358; P. W. Hodge (1971), 'Large decrease in the clear air transmission of the atmosphere 1·7 km above Los Angeles', *Nature* 229, 549; and B. C. Weare, R. L. Temkin, and F. M. Small (1974), 'Aerosols and climate: some further considerations', *Science* 186, 827–8.

General reviews on the role of man include H. E. Lansberg (1970), 'Man-made climatic changes', *Science* 170, 1265–74; M. I. Budyko *et al.* (1971), 'The impact of economic activity on climate', *Soviet Geography* 12, 666–79; and J. S. Sawyer (1971), 'Possible effects of human activity on world climate', *Weather* 26, 251–62.

There is no one good treatment of the cycle question, though H. H. Lamb (1972), *Climate, present, past, and future* vol. 1, has gone a long way towards correcting this. E. Huntington (1945) *Mainsprings of Civilization* contains a fascinating but rather one-sided commentary on the relations between climatic and economic cycles. A balanced review on the question of prediction is that of B. J. Mason (1976), 'The nature and prediction of climatic changes', *Endeavour* 35, 51–7. A more lengthy and controversial discussion of climate and its relationship to man's future is J. Gribbin's (1976), *Forecasts, famines and freezes* (Wildwood House, London). The future also concerns some of the contributors to R. J. Kopec (ed.) (1976), *Atmospheric quality and climatic change* (University of North Carolina, Chapel Hill).

References

AHLMANN, H. W. (1948) 'The present climatic fluctuation', *Geographical Journal* 112, 165–95.

AHLMANN, H. W. (1953) 'Glacier variations and climatic fluctuations', *American Geographical Society* (New York).

AHMAD, N. and SAXENA, H. B. (1963) 'Glaciations of the Pindar river valley, southern Himalayas', *Journal of Glaciology* 4, 471–6.

ANANOVA, E. N. (1967) 'Palynological correlation of the flora and vegetation of the Likhvin-Mazovian I-Holstein-Neede Interglacial', *Review of Palaeobotany and Palynology* 4, 175–86.

ANDREWS, J. T. (1970) 'A geomorphological study of Post-Glacial uplift with particular reference to Arctic Canada', *Institute of British Geographers Special Publication* 2.

ANDREWS, J. T. (1975) *Glacial systems—an approach to glaciers and their environments* (Duxbury Press, North Scituate, Mass.).

BALLARD, R. D. and UCHUPI, E. (1970) 'Morphology and Quaternary history of the continental shelf of the Gulf Coast of the United States', *Bulletin of Marine Science* 20, 547–59.

BANDY, O. L. (1968) 'Changes in Neogene paleo-oceanography and eustatic changes', *Palaeogeography, Palaeoclimatology, Palaeo-ecology* 5, 63–75.

BANDY, O. L. and MARICOVICH, L. (1973) 'Rates of Late Cenozoic uplift, Baldwin Hills, Los Angeles, California', *Science* 181, 653–4.

BARRETT, E. C. (1966) 'Regional variations of rainfall trends in northern England, 1900–1959', *Transactions, Institute of British Geographers* 38, 41–58.

BEADLE, L. C. (1974) *The inland waters of tropical Africa: an introduction to tropical limnology* (Longman, London).

BECKINSALE, R. P. (1965) 'Climatic change: a critique of modern theories', in Whittow, J. B. and Wood, P. D. (eds.), *Essays in Geography for Austin Miller* (University of Reading, Reading), 1–38.

BENEDICT, J. B. (1968) 'Recent glacial history of an Alpine area in the Colorado Front Range, U.S.A.', *Journal of Glaciology* 7, 77–87.

BERGGREN, W. A. (1969) 'Cainozoic stratigraphic, planktonic foraminiferal zonation and the radiometric time-scale', *Nature* 224, 1072–5.

BERGHINZ, C. (1976) 'Venice is sinking into the sea', in Tank, R. (ed.), *Focus on environmental geology* 2nd ed. (O.U.P., New York), 512–18.

BEVERTON, R. J. H. and LEE, A. J. (1965) 'Hydrographic fluctuations in the north Atlantic and some Biological consequences', in Johnson, C. G. and Smith, L. P. (eds.), *The biological significance of climatic changes in Britain* (Academic Press, London), 79–107.

BINNS, R. E. (1972) 'Flandrian strandline chronology for the British Isles and correlation of some European Post-glacial strandlines', *Nature* 235, 206–10.

BIRKELAND, P. W. (1972) 'Late Quaternary eustatic sea-level changes along the Malibu coast, Los Angeles County, California', *Journal of Geology* 80, 432–44.

BLOOM, A. L. (1967) Pleistocene shorelines: a new test of isostasy, *Bulletin Geological Society of America* 78, 1477–94.

BLOOM, A. L. (1971) 'Glacial eustatic and isostatic controls of sea-level since the Last Glaciation', in Turekian, K. K. (ed.), *The Late Cenozoic Glacial Ages* (Yale U.P., New Haven), pp. 355–79.

BONATTI, E. (1966) 'North Mediterranean climate during the last Würm Glaciation', *Nature* 209, 984.

BORTENSCHLAGER, S. and PATZELT, G. (1969) 'Wärmezeitliche Klima- und Gleischerswankungen im Pollin Profil Eines Hochgelegenen Moores (2270 m) Der Venediggergruppe', *Eiszeitalter und Gegenwart* 20, 116–22.

BOWEN, A. J. (1972) 'The tidal régime of the River Thames; long-term trends and their possible causes', *Philosophical Transactions Royal Society of London* A, 272, 187–99.

BOWEN, D. Q. (1970) 'South-east and central South Wales', in Lewis C. A. (ed.), *The glaciations of Wales and adjoining regions* (Longman, London), pp. 197–227.

BOWEN, D. Q. (1973) 'The Pleistocene history of Wales and the Borderland', *Geological Journal* 8 (2), 207–24.

BOWLER, J. M. (1976) 'Aridity in Australia: age, origins, and expression in aeolian landforms and sediments', *Earth Science Review* 12, 279–310.

BOWLES, F. A. (1976) 'Paleoclimatic significance of quartz/illite variations in cores from the eastern equatorial North Atlantic', *Quaternary Research* 5, 225–35.

BRADLEY, R. S. and MILLER, G. H. (1972) 'Recent climatic change and increased glacierization in the eastern Canadian Arctic', *Nature* 237, 385–7.

BRAUDEL, F. (1972) *The Mediterranean and the Mediterranean world in the age of Philip II* (Collins, London), vol. 1. pp. 642ff.

BRAY, J. R. (1970) 'Temporal patterning of post-Pleistocene glaciation', *Nature* 228, 353–4.

BRAY, J. R. (1974) 'Volcanism and glaciation during the past 40 millennia', *Nature* 252, 679–80.

BRINK, N. W. T. and WEIDICK, A. (1974) 'Greenland ice sheet history since the Last Glaciation', *Quaternary Research* 4, 429–40.

BROECKER, W. S. (1975) 'Climatic change: are we on the brink of a pronounced global warming?', *Science* 189, 460–3.

BROECKER, W. S. and KAUFMANN, A. (1965) 'Radiocarbon chronology of Lake Lahontan and Lake Bonneville II, Great Basin', *Bulletin Geological Society of America* 76, 537–66.

BROECKER, W. S. and THURBER, D. L. (1965) 'Uranium Series dating of corals and oolites from Bahaman and Florida Keys Limestones', *Science* 149, 58–60.

BROECKER, W. S., THURBER, D. L., GODDARD, J., KU, T. L., MATTHEWS, R. K., and MESOLLELA, K. J. (1968) 'Milankovitch hypothesis supported by precise dating of coral reefs and deep-sea sediments', *Science* 159, 297–300.

BROOKS, C. E. P. (1926) *Climate through the ages* (Benn, London).

BROWN, J. A. (1976) 'Shortening of growing season in the U.S. Corn Belt', *Nature* 260, 420–1.

BROWH, K. S., SHEPPARD, P. M., and TURNER, J. R. G. (1974) 'Quaternary refugia in tropical America: evidence from race formation in *Heliconius* Butterflies', *Proceedings, Royal Society* B, 369–78.

BROWN, P. R. (1953) 'Climatic fluctuation in the Greenland and Norwegian Seas', *Quarterly Journal of the Royal Meteorological Society* 79, 272–81.

BRYAN, K. (1941) 'Pre-Columbian agriculture in the south-west as conditioned by periods of alluviation', *Annals Association of American Geographers* 31, 219–42.

BRYSON, R. A. (1974) 'A perspective on climatic change', *Science* 184, 753–60.

BRYSON, R. A. and BARREIS, D. A. (1967) 'Possibility of major climatic modification and their implications: Northwest India, A case for study', *Bulletin American Meteorological Society* 48, 136–42.

BUDYKO, M. A., DROZDOV, D. A., and YUDIN, M. I. (1971) 'The impact of economic activity on climate', *Soviet Geography* 12, 666–79.

BUNTING, A. H., DENNETT, M. D., ELSTON, J., and MILFORD, J. R. (1976) 'Rainfall trends in the West African Sahel', *Quarterly Journal of the Royal Meteorological Society* 102, 59–64.

BUTZER, K. W. (1961) 'Climatic change in arid regions since the Pliocene', *Arid Zone Research* (UNESCO) 17, 31–56.

BUTZER, K. W. (1971) 'Recent history of an Ethiopian delta', *University of Chicago, Dept of Geography, Research Paper* 136, p. 184.

BUTZER, K. W. (1972) *Environment and archeology—an ecological approach to prehistory* (Methuen, London), pp. 703ff.

BUTZER, K. W. (1975) 'Geological and ecological perspectives on the Middle Pleistocene', in Butzer, K. W. and Isaac, G. L., *After the Australopithecines* (Mouton, The Hague), 857–73.

BUTZER, K. W. and HANSEN, C. L. (1968) *Desert and river in Nubia: geomorphology and prehistoric environments at the Aswan Reservoir* (U. of Wisconsin Press).

BUTZER, K. W., ISAAC, G. L., RICHARDSON, J. L., and WASHBOURN-KAMAU, C. (1972) 'Radiocarbon dating of East African lake levels', *Science* 175, 1069–75.

CALDER, N. (1974) *The weather machine* (B.B.C., London).

CALLENDAR, G. S. (1961) 'Temperature fluctuations and trends over the Earth', *Quarterly Journal of the Royal Meteorological Society* 87, 1–12.

CHALINE, J. (1972) *Le Quaternaire* (Doin, Paris).

CHARLESWORTH, J. K. (1957) *The Quaternary Era* (Arnolds, London), pp. 1700ff.

CHARNEY, J., STONE, P. H., and QUIRK, W. J. (1975) 'Drought in the Sahara: a biogeophysical feedback mechanism', *Science* 187, 434–5.

CHILDE, V. G. (1954) *New light on the most ancient east* (Routledge Kegan & Paul, London).

CHU KO-CHAN (1973) 'A preliminary study on the climatic fluctuations during the last 5000 years in China', *Scientia Sinica* 16, 226–56.

CLAIBORNE, R. (1973) *Climate, man and history* (Angus and Robertson, London), pp. 444ff.

CLARK, J. A. (1976) 'Greenland's rapid post-glacial emergence: a result of ice-water gravitational attraction', *Geology* 4, 310–12.

CLARK, J. D. (1975) 'Africa in prehistory: peripheral or paramount?', *Man* 10, 175–98.

CLARK, J. G. D. (1970) 'Mesolithic times', in *Cambridge ancient history* 3rd edn., vol. 1, 90–121.

CLARKE, R. H. (1970) 'Quaternary sediments off south-east Devon', *Quarterly Journal of the Geological Society* 125 (3), 277–318.

CLIMAP PROJECT MEMBERS (1976) 'The surface of the Ice-Age earth', *Science* 191, 1131–7.

COETZEE, J. A. (1964) 'Evidence for a considerable depression of the vegetation belts during the Upper Pleistocene on the East African mountains', *Nature* 204, 564–6.

COLEMAN, J. M. (1964) 'Late recent rise of sea-level', *Bulletin Geological Society of America* 75, 833–40.

COLINVAUX, P. A. (1972) 'Climate and the Galapagos Islands', *Nature* 240, 17–20.

COLLINS, B. W. and FRASER, R. (eds.) (1971) 'Recent crustal movements', *Bulletin Royal Society of New Zealand* No. 9.

CONOVER, J. H. (1967) 'Are New England winters getting milder?—II', *Weatherwise* 20, 58–61.

COOK, P. J. and POLACH, H. A. (1973) 'A Chenier sequence at Broad Sound, Queensland, and evidence against a Holocene high sea-level', *Marine Geology* 14, 253–68.

COOKE, H. B. S. (1973) 'Pleistocene chronology: long or short?', *Quaternary Research* 3, 206–20.

COOKE, H. J. (1975) 'The Palaeoclimatic significance of caves and adjacent landforms in western Ngamiland, Botswana', *Geographical Journal* 141, 430–44.

COOKE, R. U. and REEVES, R. W. (1976) *Arroyos and environmental change in the American south-west* (Clarendon Press, Oxford), pp. 213ff.

COOPE, G. R. (1975) 'Climatic fluctuations in northwest Europe since the Last Interglacial, indicated by fossil assemblages of Coleoptera, in Wright, A. E. and Moseley, F. (eds.), *Ice ages: ancient and modern* (Seel House, Liverpool), 153–68.

COOPE, G. R. (1975) 'Mid-Weichselian climatic changes in western Europe reinterpreted from Coleopteran assemblages', *Bulletin Royal Society of New Zealand* 13, 101–8.

COOPE, G. R., MORGAN, A., and OSBORNE, P. J. (1971) 'Fossil Coleoptera as indicators of climatic fluctuations during the Last Glaciation in Britain', *Palaeogeography, Palaeoclimatology, Palaeoecology* 10, 87–101.

COX, A., DOELL, R. R., and DALRYMPLE, G. B. (1968) 'Radiometric time-scale for geomagnetic reversals', *Quarterly Journal of the Geological Society* 124, 53–66.

COX, A. (1968) 'Polar wandering, continental drift and the onset of Quaternary glaciation', *Meteorological Monographs* 8, 112–25.

CRISP, D. J. (1959) 'The influence of climatic changes on animals and plants', *Geographical Journal* 125, 1–19.

CRITTENDEN, M. D. (1963) 'New data on the isostatic deformation of Lake Bonneville', *United States Geological Survey Professional Paper* 454–E, pp. 31ff.

CURRY, R. R. (1969) 'Holocene climatic and glacial history of the Central Sierra Nevada', *Geological Society of America Special Paper* 123, 1–47.

CUSHING, D. (1976) 'The impact of climatic change on fish stocks in the North Atlantic', *Geographical Journal* 142, 216–27.

DALY, R. A. (1926) *Our mobile Earth* (Scribners, New York), pp. 342ff.

DAMUTH, J. E. and FAIRBRIDGE, R. W. (1970) 'Arkosic sands of the Last Glacial stage in the tropical Atlantic off Brazil', *Bulletin Geological Society of America* 81, 189–206.

DANSGAARD, W. (1969) 'One thousand centuries of climatic record from Camp Century on the Greenland ice sheet', *Science*, 166, 377–80.

DANSGAARD, W., JOHNSEN, S. J., and CLAUSEN, H. B. *et al.* (1970) 'Ice cores and paleoclimatology', in Olsson, I. U. (ed.) *Radiocarbon Variation and absolute chronology* (Wiley, New York), 337–51.

DANSGAARD, W., JOHNSEN, S. J., CLAUSEN, H. B., and LANGWAY, C. C. (1971) 'Climatic record revealed by the Camp Century Ice Core', in Turekian, K. K. (ed.), *The Late Cenozoic glacial ages* (Yale University Press, New Haven), 37–56.

DANSGAARD, W., JOHNSEN, S. J., REEH, N., OUNDERSTRUP, N., CLAUSEN, H. B., and HAMMER, C. U. (1975) 'Climatic changes, Norsemen and modern man', *Nature* 255, 24.

DARWIN, C. (1936) *Origin of species* (Minerva, London).

DAVIS, N. E. (1972) 'The variability of the onset of spring in Britain', *Quarterly Journal of the Royal Meteorological Society* 98, 763–77.

DAVITAYA, F. F. (1969) 'Atmospheric dust content as a factor affecting glaciation and climatic change', *Annals, Association of American Geographers* 59, 552–60.

DEEVEY, E. S. (1949) 'Biogeography of the Pleistocene', *Bulletin Geological Society of America* 60, 1315–1416.

DEEVEY, E. S. and FLINT, R. F. (1957) 'Post-Glacial hypsithermal interval', *Science* 125, 182–4.

DEGENS, E. T. and HECKY, R. E. (1974) 'Paleoclimatic reconstruction of Late

Pleistocene and Holocene based on biogenic sediments from the Black Sea and a tropical African lake', *Colloques Internationaux du CNRS* 219, 13–23.

DE JONG, J. D. (1967) 'The Quaternary of the Netherlands', in Rankama, K. (ed.), *The Quaternary* vol. 2, 301–426.

DEMOUGEOT, E. (1965) 'Variations climatiques et invasions', *Rev. Hist.* 228, 1–22.

DENTON, G. H. and KARLEN, W. (1973) 'Holocene climatic variations: their pattern and possible cause', *Quaternary Research* 3, 155–205.

DENTON, G. H. and PORTER, S. C. (1970) 'Neo-glaciation', *Scientific American* 222 (6) 101–10.

DE PLOEY, J. (1965) 'Position Géomorphologique, Génèse et Chronologie de Certains Dépôts Superficiels au Congo Occidental', *Quaternaria* 7, 131–54.

DEUSER, W. G., ROSS, E. H., and WATERMAN, L. S. (1976) 'Glacial and pluvial periods: their relationship revealed by Pleistocene sediments of the Red Sea and Gulf of Aden', *Science* 191, 1168–70.

DIAMOND, M. (1958) 'Precipitation trends in Greenland during past 30 years', *Journal of Glaciology* 3, 177–80.

DONN, W. L., FARRAND, W. R., and EWING, M. (1962) 'Pleistocene ice volumes and sea-level changes', *Journal of Geology* 206–14.

DONN, W. L. and SHAW, D. M. (1963) 'Sea-level and climate of the past century', *Science* 142, 1166–7.

DONNER, J. J. (1970) 'Land/sea-level changes in Scotland', in Walker, D. and West, R. G. (eds.), *Studies in the vegetational history of the British Isles* (Cambridge University Press, Cambridge), 23–39.

DREIMANIS, A. and RAUKAS, A. (1975) 'Did Middle Wisconsin, Middle Weichselian, and their equivalents represent an interglacial or an interstadial complex in the Northern Hemisphere?', *Bulletin Royal Society of New Zealand* 13, 109–20.

DREIMANIS, A., TERASMAE, J., and McKENZIE, G. D. (1966) 'The Port Talbot Interstade of the Wisconsin Glaciation', *Canadian Journal of Earth Sciences* 3, 305–25.

DREWRY, D. J. (1975) 'Initiation and growth of the east Antarctic ice sheet', *Journal Geological Society of London* 131, 255–73.

DUNN, G. E. and MILLER, B. I. (1960) *Atlantic hurricanes* (Louisiana State University Press), pp. 326ff.

DUPLESSY, J. C., LABEYRIE, J., LALOU, C., and NGUYEN, H. V. (1970) 'Continental climatic variations between 130 000 and 90 000 Years B.P.', *Nature* 226, 631–3.

DURHAM, J. W. (1950) 'Cenozoic marine climates of the Pacific coast', *Bulletin Geological Society of America* 61, 1243–64.

DURY, G. H. (1965) 'Theoretical implications of underfit streams', *United States Geological Survey Professional Paper* 452C, C1–C43.

DURY, G. H. (1967) 'Climatic change as a geographical backdrop', *Australian Geographer* 10, 231–42.

DURY, G. H. (1973) 'Paleo-hydrologic implications of some pluvial lakes in Northwestern New South Wales, Australia', *Bulletin, Geological Society of America* 84, 3663–76.

EARDLEY, A. J. and GVOSDETSKY, V. (1960) 'Analysis of Pleistocene core from Great Salt Lake, Utah', *Bulletin Geological Society of America* 71, 1323–44.

EAST, W. G. (1938) *The geography behind history* (Nelson, London), pp. 200ff.

EMBLETON, C. and KING, C. A. M. (1967) *Glacial and periglacial geomorphology* (Arnold's, London).

EMERY, K. O. (1969) 'The continental shelves', *Scientific American* 221, 107–22.

EMERY, K. O., NIINO, H., and SULLIVAN, B. (1971) 'Post Pleistocene levels of the East China Sea', in Turekian, K. K. (ed.), *The Late Cenozoic glacial ages* (Yale University Press, New Haven), 381–90.

EMILIANI, C. (1961) 'Cenozoic climatic changes as indicated by the stratigraphy and chronology of deep-sea cores of Globigerina-Ooze facies', *Annals New York Academy of Science* 95, 521–36.

EMILIANI, C. (1966) 'Isotopic paleotemperatures', *Science* 154, 851–7.

EMILIANI, C. (1966) 'Palaeotemperature analysis of Caribbean cores P 6304–8 and P 6304–9 and a generalized temperature curve for the past 425 000 years', *Journal of Geology* 74, 109–26.

EMILIANI, C. (1968) 'The Pleistocene Epoch and the evolution of man', *Current Anthropology* 9, 27–47.

EMILIANI, C. and FLINT, R. F. (1963) 'The Pleistocene record', in Hill, M. N. (ed.), *The Sea* vol. 3 (Wiley, New York), pp. 888–927.

EPSTEIN, S., SHARP, R. P., and GOW, A. J. (1970) 'Antarctic ice sheet: stable isotope analysis of Byrd Station cores and interhemispheric climatic implications', *Science* 168, 1570–2.

ERICSSON, D. B., EWIN, M., WOLLIN, G., and HEEZEN, B. C. (1961) 'Atlantic deep-sea sediment cores', *Bulletin Geological Society of America* 72, 193–286.

ERICSON, D. B. and WOLLIN, G. (1968) 'Pleistocene climates and chronology in deep-sea sediments', *Science* 162, 1227–34.

EVANS, P. (1971) 'Towards a Pleistocene time-scale', in *The Phanerozoic time-scale—A supplement* (Geological Society, London), Part 2, 123–356.

EWING, M. (1971) 'The Late Cenozoic history of the Atlantic Basin, and its bearing on the cause of the Ice Ages', in Turekian, K. K. (ed.), *The Late Cenozoic glacial ages* (Yale U.P., New Haven), pp. 565–73.

EWING, M. and DONN, W. L. (1956) 'A theory of ice ages', *Science* 123, 1061–6.

EWING, M. and DONN, W. L. (1958) 'A theory of ice ages', *Science* 127, 1159–62.

FAIRBRIDGE, R. W. (1958) 'Dating the latest movements of the Quaternary sea-level', *Transactions, New York Academy of Sciences* 20, 471–82.

FAIRBRIDGE, R. W. (1961) 'Eustatic changes in sea-level', *Physics and Chemistry of The Earth* 4, 99–185.

FAIRBRIDGE, R. W. and KREBS, O. A. (1962) 'Sea-level and the southern oscillation', *Geophysics Journal* 6, 532–45.

FARRAND, W. R. (1971) 'Late Quaternary paleoclimates of the eastern Mediterranean area', in Turekian, K. K. (ed.), *The Late Cenozoic glacial ages* (Yale U.P., New Haven), pp. 529–64.

FAURE, H. (1966) 'Evolution des Grands Lacs Sahariens a l'Holocene', *Quaternaria* 8, 167–75.

FIELD, W. O. (1954) 'Notes on the advance of Taku Glacier', *Geographical Review* 44, 236–9.

FLEMMING, N. C. (1969) 'Archaeological evidence for eustatic change of sea-level and earth movements in the western Mediterranean during the last 2000 years', *Geological Society of America Special Paper* 109.

FLEMMING, N. C. (1972) 'Relative chronology of submerged Pleistocene marine erosion features in the western Mediterranean', *Journal of Geology* 80, 633–62.

FLINT, R. F. (1971) *Glacial and Quaternary geology* (Wiley, New York).

FLINT, R. F. and BOND, G. (1968) 'Pleistocene sand ridges and pans in western Rhodesia', *Bulletin Geological Society of America* 79, 299–313.

FLORSCHÜTZ, F., MENENDEZ AMOR, J., and WIJMSTRA, T. A. (1974) 'Palynology of a thick Quaternary succession in southern Spain', *Palaeogeography, Palaeoclimatology Palaeo-ecology* 10, 233–64.

FONTES, J. CH. and BORTOLAMI, G. (1973) 'Subsidence of the Venice area during the past 40 000 years', *Nature* 244, 339–41.

FRANKLIN, J. F., NOIR, W. H., DOUGLAS, G. W., and WIBERG, C. (1971) 'Invasion of sub-alpine meadows by trees in the Cascade Range, Washington and Oregon', *Arctic and Alpine Research* 3, 215–24.

FRENZEL, B. (1973) *Climatic fluctuations of the Ice Age* (Case Western Reserve University, Cleveland), pp. 300ff.

FUJITA, T. T. (1973) 'Tornadoes round the world', *Weatherwise* 26.

FUKUI, E. (1970) 'The recent rise of temperature in Japan', *Japanese Progress in Climatology*, 46–55.

FULTON, R. J. (1968) 'Olympia Interglaciation, Purcell Trench, British Columbia', *Bulletin Geological Society of America* 79, 1075–80.

FUNDER, S. (1972) 'Deglaciation of the Scoresby Sund Fjord region, north-east Greenland', in Price, R. J. and Sugden, D. E. (eds.), *Polar geomorphology* (I.B.G. Special Publication 4), 33–41.

GALLOWAY, R. W. (1970) 'The full glacial climate in South-Western U.S.A.', *Annals Association of American Geographers* 60, 245–56.

GAMBOLATI, G., GATTO, P., and FREEZE, R. A. (1974) 'Predictive simulation of the subsidence of Venice', *Science* 183, 849–51.

GATES, W. L. (1976) 'Modeling the Ice-Age climate', *Science* 191, 1138–44.

GEITZENAUER, K. R., MARGOLIS, S. V., and EDWARDS, D. S. (1968) 'Evidence consistent with Eocene glaciation in a south Pacific deep-sea sedimentary core', *Earth and Planetary Science Letters* 4, 173–7.

GENTILLI, J. (1961) 'Quaternary climates of the Australian region', *Annals New York Academy of Sciences* 95, 465–501.

GENTILLI, J. (1971) *Climates of Australia and New Guinea* (Elsevier, Amsterdam), pp. 405ff.

GERASIMOV, I. B. (1969) 'Degradation of the last European ice sheet', in Wright, H. E. (ed.), *Quaternary geology and climate* (National Academy of Sciences, Washington), 72–8.

GILL, E. D. (1961) 'Cainozoic climates of Australia', *Annals New York Academy of Sciences* 95, 461–4.

GJAEVEROLL, O. (1963) 'Survival of plants on nunataks in Norway during the Pleistocene glaciation', in Löve A. and Löve, D. *North Atlantic biota and their history* (Pergamon, Oxford), 261–83.

GLASS, B., ERICSON, D. B., HEEZEN, B. C., OPDYKE, N. D., and GLASS, J. A. (1967) 'Geomagnetic reversals and Pleistocene chronology', *Nature* 216, 437–42.

GLASSFORD, D. K. and KILLIGREW, L. P. (1976) 'Evidence for Quaternary westward extension of the Australian Desert into south-western Australia', *Search* 7, 394–96.

GODWIN, H. (1956) *The history of the British flora* (Cambridge University Press, Cambridge).

GODWIN, H., SUGGATE, R. P., and WILLIS, E. H. (1958) 'Radiocarbon dating of the eustatic rise in ocean level', *Nature* 181, 1518–19.

GOLDTHWAIT, R. P., MCKELLAR, I. C., and CRONK, C. (1963) 'Fluctuations of the Crillon Glacier system, south east Alaska', *Bulletin International Association of Scientific Hydrology* 8, 62–74.

GOUDIE, A. S. (1972) 'The concept of Post-Glacial progressive desiccation', *Research Paper* No. 4 (School of Geography, University of Oxford).

GOUDIE, A. S., ALLCHIN, B., and HEGDE, K. T. M. (1973) 'The former extensions of the Great Indian Sand Desert', *Geographical Journal* 139, 243–57.

GOW, A. J. and WILLIAMSON, T. (1971) 'Volcanic ash in the Antarctic ice sheet and its possible climatic implications', *Earth and Planetary Science Letters* 13, 210–18.

GRASTY, R. L. (1967) 'Orogeny, a cause of world-wide regression of the seas', *Nature* 216, 779.

GREGORY, S. (1956) 'Regional variations in the trend of annual rainfall over the British Isles', *Geographical Journal* 122, 346–53.

GREEN, C. P. (1973) 'Pleistocene river gravels and the Stonehenge problem', *Nature* 243, 214–16.

GREEN, C. P. (1974) 'Pleistocene gravels of the river Axe in south-western England, and their bearing on the southern limit of glaciation in Britain', *Geological Magazine* 111, 213–20.

GRIBBIN, J. (1975) 'Aerosol and climate: hotter or cooler', *Nature* 253, 162.

GRIFFIN, J. B. (1967) 'Climatic change in American prehistory', in Fairbridge, R. W. (ed.), *The Encyclopedia of atmospheric sciences and astrogeology* 169–71.

GROVE, A. T. (1958) 'The ancient Erg of Hausaland, and similar formations on the south side of the Sahara', *Geographical Journal* 124, 528–33.

GROVE, A. T. (1967) *Africa south of the Sahara* (O.U.P., London).

GROVE, A. T. (1969) 'Landforms and climatic change in the Kalahari and Ngamiland', *Geographical Journal* 135, 192–212.

GROVE, A. T. and WARREN, A. (1968) 'Quaternary landforms and climate on the south side of the Sahara', *Geographical Journal* 134, 194–208.

GROVE, A. T., STREET, F. A., and GOUDIE, A. S. (1975) 'Former lake levels and climatic change in the Rift valley of southern Ethiopia', *Geographical Journal* 141, 177–94.

GROVE, J. M. (1966) 'The Little Ice Age in the Massif of Mont Blanc', *Transactions Institute of British Geographers* 40, 129–43.

GROVE, J. M. (1972) 'The incidence of landslides, avalanches and floods in western Norway during the Little Ice Age', *Arctic and Alpine Research* 4, 131–8.

GUILCHER, A. (1969) 'Pleistocene and Holocene sea-level changes', *Earth Science Reviews* 5, 69–98.

GUILDAY, J. E. (1967) 'Differential extinction during late Pleistocene and Recent times', in Martin, P. S. and Wright, H. E. (eds.), *Pleistocene extinctions* 121–40 (Yale University Press, New Haven).

GUTENBERG, B. (1941) 'Changes in sea-level, post-glacial uplift, and mobility of the Earth's interior, *Bulletin Geological Society of America* 52, 721–72.

HACK, J. T. N. (1941) 'Dunes of the western Navajo Country', *Geographical Review* 31, 240–63.

HAFFER, J. (1969) 'Speciation in Amazonian forest birds', *Science* 165, 131–7.

HAMMEN, T. VAN DER (1972) 'Changes in vegetation and climate in the Amazon basin and surrounding areas during the Pleistocene', *Geologie en Mijnbouw* 51, 641–3.

HAMMEN, T. VAN DER (1974) 'The Pleistocene changes of vegetation and climate in tropical South America', *Journal of Biogeography* 1, 3–26.

HAMMEN, T. VAN DER, MAARLEVELD, G. C., VOGEL, J. C., and ZAGWIJN, W. H. (1967) 'Stratigraphy, climatic succession, and radiocarbon dating of the Last Glacial in the Netherlands', *Geologie en Mijnbouw* 46, 79–95.

HAMMEN, T. VAN DER, WIJMSTRA, T. A., and ZAGWIJN, W. H. (1971) 'The floral record of the Late Cenozoic of Europe', in Turekian, K. K. (ed.), *The Late Cenozoic glacial ages* 391–424 (Yale University Press, New Haven).

HARRIS, G. (1964) 'Climatic changes since 1860 affecting European birds', *Weather* 19, 70–9.

HAWKINS, A. B. (1971) 'The Late Weichselian and Flandrian transgression of south west Britain', *Quaternaria* 14, 115–30.

HECHT, A. D. (1974) 'Quantitative micropaleontology and the amplitude of glacial interglacial temperature changes in the Caribbean Sea, Gulf of Mexico, and equatorial Atlantic', *Colloques Internationaux du CNRS* 219, 213–20.

HENDY, C. H. and WILSON, A. T. (1968) 'Palaeoclimate data from speleothems', *Nature* 219, 48–51.

HERMANN, Y. (1970) 'Arctic palaeo-oceanography in Late Cenozoic time', *Science*, 169, 474–7.

HEUSSER, C. J. (1961) 'Some comparisons between climatic changes in north-western North America and Patagonia', *Annals New York Academy of Sciences* 95, 642–57.

HEUSSER, C. J., SCHUSTER, R. L., and GILKEY, A. K. (1954) 'Geobotanical studies on the Taku Glacier anomaly', *Geographical Review* 44, 224–36.

HEUSSER, C. J. and MARCUS, M. G. (1964) 'Historical variations of Lemon Creek Glacier, Alaska, and their relationship to the climatic history', *Journal of Glaciology* 5, 77–86.

HEY, R. W. (1963) 'Pleistocene screes in Cyrenaica (Libya)', *Eiszeitalter und Gegenwart* 14, 77–84.

HIGGINS, C. G. (1965) 'Causes of relative sea-level changes', *American Scientist* 53, 464–76.

HIGGINS, C. G. (1969) 'Isostatic effects of sea-level changes', in Wright, H. E. (ed.), *Quaternary geology and climate* (National Academy of Sciences, Washington), pp. 141–5.

HOFFMAN, R. S. and JONES, J. K. (1970) 'Influence of Late Glacial and Post-Glacial events on the distribution of recent mammals on the northern Great Plains', in Dort, W. and Jones, J. K. (eds.), *Pleistocene and recent environments of the central Great Plains* (Kansas U.P.), 355–94.

HOINKES, H. C. (1968) 'Glacier variation and weather', *Journal of Glaciology* 7, 3–20.

HOLLIN, J. T. (1965) 'Wilson's theory of ice ages', *Nature* 208, 12–16.

HORNER, R. W. (1972) 'Current proposals for the Thames Barrier and the organization of the investigations', *Philosophical Transactions Royal Society of London* A, 272, 179–85.

HOWE, G. M., SLAYMAKER, H. O., and HARDING, D. M. (1966) 'Flood hazard in Mid-Wales', *Nature* 212, 584–5.

HSIEH, CHIAO-MIN (1976) 'Chu K'O-Chen and China's climatic change', *Geographical Journal* 142, 248–56.

HUNTINGTON, E. (1945) *Mainsprings of civilisation* (Wiley, New York).

IRWIN-WILLIAMS, C. and HAYNES, C. V. (1970) 'Climatic change and early population dynamics in the south-western United States', *Quaternary Research* 1, 59–71.

ISAAC, E. (1970) *Geography of domestication* (Prentice Hall, Englewood Cliffs), pp. 132ff.

JELGERSMA, S. (1966) 'Sea-level changes in the last 10 000 years', in *Royal Meteorological Society International Symposium on World Climate From 8000–O B.C.* 54–69.

JETT, S. C. (1964) 'Pueblo Indian migrations: an evaluation of the possible physical and cultural determinants', *American Antiquity* 29, 281–300.

JOHNSEN, S. J., DANSGAARD, W., CLAUSEN, H. B., and LANGWAY, C. C. (1970) 'Climatic oscillations 1200–2000 A.D.', *Nature* 227, 482.

KAFRI, U. (1969) 'Recent crustal movements in northern Israel', *Journal of Geophysics Research* 74, 4246–58.

KAISER, K. (1969) 'The climate of Europe during the Quaternary Ice Age', in Wright, H. E. (ed.), *Quaternary Geology and Climate* (National Academy of Science, Washington), 10–37.

KALELA, O. (1952) 'Changes in the geographic distribution of Finnish birds and mammals in relation to recent changes in climate', *Fennia* 75, 38–51.

KALNICKY, R. A. (1974) 'Climatic change since 1950', *Annals Association of American Geographers* 64, 100–12.

KARLÉN, W. (1973) 'Holocene glacier and climatic variations, Kebnekaise Mountains, Swedish Lapland', *Geografiska Annaler* 55(A), 29–63.

KAYE, C. A. and BARGHOORN, E. (1964) 'Late Quaternary sea-level rise at Boston, Mass., and notes on the autocompaction of peat', *Bulletin Geological Society of America* 75, 63–80.

KEANY, J., LEDBETTER, M., WATKINS, N., and TER CHIEN H. (1976) 'Diachronous deposition of ice-rafted debris in sub-Antarctic deep-sea sediments', *Bulletin Geological Society of America* 87, 873–82.

KELLAWAY, G. A. (1971) 'Glaciation and the stones of Stonehenge', *Nature* 233, 30–5.

KENDALL, R. L. (1969) 'An ecological history of the Lake Victoria Basin', *Ecological Monographs* 39, 121–76.

KENNETT, J. P. (1970) 'Pleistocene paleoclimates and foraminiferal biostratigraphy in Sub-Antarctic deep-sea cores', *Deep-Sea Research* 17, 125–40.

KENNETT, J. P. and SHACKLETON, N. J. (1975) 'Laurentide ice sheet meltwater recorded in Gulf of Mexico deep-sea cores', *Science* 188, 147–50.

KENNETT, J. P. and THUNELL, R. C. (1975) 'Global increase in Quaternary explosive volcanism', *Science* 187, 497–503.

KENT, D., OPDYKE, N. D., and EWING, M. (1971) 'Climate change in the North Pacific using ice-rafted detritus as a climatic indicator', *Bulletin Geological Society of America* 82, 2741–54.

KERANEN, J. (1952) 'On temperature changes in Finland during the last 100 years', *Fennia* 75, 5–16.

KERSHAW, A. P. (1974) 'A long continuous pollen sequence from north-eastern Australia', *Nature* 251, 222.

KIDSON, C. (1977) 'The coast of south west England' in Kidson, C. and Tooley, M. J. (eds.), *The Quarternary history of the Irish Sea*, 257–298.

KIND, N. V. (1972) 'Late Quaternary climatic changes and glacial events in the Old and New World—radiocarbon chronology', *24th International Geographical Congress* Section 12, 55–61.

KING, J. W. (1973) 'Solar radiation changes and the weather', *Nature* 245, 443–6.

KING, L. C. (1962) *Morphology of the Earth* (Oliver and Boyd, Edinburgh).

KING, P. B. (1965) 'Tectonics of Quaternary time in middle North America', in Wright, H. E. and Frey, D. G. (eds.), *The Quaternary of the U.S.A.*, 831–70.

KLEIN, C. (1965) 'On the fluctuations of the level of the Dead Sea since the beginning of the nineteenth century', *State of Israel, Hydrological Service, Hydrological Paper* 7

KOSIBA, A. (1963) 'Changes in the Werenskiold Glacier and Hans Glacier in SW. Spitzbergen', *Bulletin International Association of Scientific Hydrology* 8, 24–35.

KRANTZ, G. S. (1970) 'Human activities and megafaunal extinctions', *American Scientist* 58, 164–70.

KRAUS, E. B. (1954) 'Secular changes in the rainfall regime of south east Australia', *Quarterly Journal of the Royal Meteorological Society* 80, 591–611.

KRAUS, E. B. (1955a) 'Secular variations of east coast rainfall regimes', *Quarterly Journal of the Royal Meteorological Society* 81, 430–9.

KRAUS, E. B. (1955b) 'Secular changes of tropical rainfall regimes', *Quarterly Journal of the Royal Meteorological Society* 81, 198–210.

KRAUS, E. B. (1956) 'Secular changes of the standing circulation', *Quarterly Journal of the Royal Meteorological Society* 82, 289–300.

KU, T. L. (1968) 'Protactinium 231 methods of dating coral from Barbados Island', *Journal of Geophysics Research*, 73, 2271–6.

KUKLA, G. J. (1975) 'Loess stratigraphy of central Europe', in Butzer, K. W. and Isaac, G. L. (eds.), *After the Australopithecines* (Mouton, The Hague), 99–188.

KURTÉN, B. (1972) *The Ice Age* (International Book Production, Stockholm).

KVITOVIC, J. and VANKO, J. (1971) 'Studium sucasnych pohybov zemskej Kory NA Slovensku', *Geograficky Casopis* 23, 124–32.

LADURIE, E. LE R. (1971) *Times of feast, times of famine* (Allen and Unwin, London).

LAMB, H. H. (1966) *The changing climate* (Methuen, London).

LAMB, H. H. (1966) 'Climate in the 1960s with special reference to East African lakes', *Geographical Journal* 132, 183–212.

LAMB, H. H. (1967) 'Britain's changing climate', *Geographical Journal* 133, 445–68.

LAMB, H. H. (1969) 'Climatic fluctuations', in Flohn, H. (ed.), *World survey of climatology*, vol. 2 (Elsevier, Amsterdam), 173–249.

LAMB, H. H. (1969) 'The new look of climatology', *Nature* 223, 1209–15.

LAMB, H. H. (1970) 'Volcanic dust in the atmosphere: with a chronology and assessment of its meteorological significance', *Philosophical Transactions, Royal Society of London*, A, 266, 425–533.

LAMB, H. H. (1971) 'Volcanic activity and climate', *Palaeogeography Palaeoclimatology Palaeo-ecology* 10, 203–30.

LAMB, H. H. (1972) *Climate: present, past and future* (Methuen, London), vol. 1, pp. 613ff.

LAMB, H. H. (1974) The current trend of world climate', *Climatic Research Unit Research Publication* 3.

LAMB, H. H., PROBERT-JONES, J. R., and SHEARD, J. W. (1962) 'A new advance of the Jan Mayen glaciers and a remarkable increase of precipitation', *Journal of Glaciology* 4, 355–65.

LANDSBERG, H. E. (1976) 'Whence global climate: hot or cold? an essay review', *Bulletin American Meteorological Society* 57, 441–3.

LARSEN, J. A. and BARRY, R. G. (1974) 'Palaeoclimatology', in Ives, J. D. and Barry, R. G., (eds.) *Arctic and alpine environments* (Methuen, London).

LAWRENCE, D. B. (1950) 'Glacier fluctuation for six centuries in south-eastern Alaska and its relation to solar activity', *Geographical Review* 40, 191–223.

LAWRENCE, D. B. (1958) 'Glaciers and vegetation in south east Alaska', *American Scientist* 46, 89–122.

LAWRENCE, D. B. and LAWRENCE, E. G. (1961) 'Response of enclosed lakes to current glacio-pluvial climatic conditions in middle-latitude western North America', *Annals New York Academy of Sciences* 95, 341–50.

LEAKEY, L. S. B. and GOODALL, V. M. (1969) *Unveiling man's origins* (Methuen, London).

LEHMER, D. J. (1970) 'Climate and culture history in the middle Missouri Valley', in Dort, W. and Jones, J. K. (eds.), *Pleistocene and recent environments of the central Great Plains* (Kansas U.P.), pp. 117–29.

LEOPOLD, L. B. (1951) 'Rainfall frequency: an aspect of climatic variation', *Transactions American Geophysics Union* 32, 347–57.

LEOPOLD, L. B., LEOPOLD, E. B., and WENDORF, F. (1963) 'Some climatic indicators in the period A.D. 1200–1400 in New Mexico', *Arid Zone Research* 20, 265–70.

LEOPOLD, L. B., WOLMAN, M. G., and MILLER, J. P. (1964) *Fluvial processes in geomorphology* (Freeman, San Francisco), pp. 522ff.

LEROI-GOURHAN, A. (1974) 'Analyses Polliniques, Pré-Histoire et Variations Climatiques Quaternaires', *Colloques Internationaux du CRNS* 219, 61–6.

LILJEQUIST, G. H. (1943) 'The severity of the winters at Stockholm 1757–1942', *Geografiska Annaler* 25, 81–97.

LISITZIN, E. (1964) 'Land uplift as sea-level problem', *Fennia* 89, 7–10.

LIVINGSTONE, D. A. (1967) 'Post-glacial vegetation of the Ruwenzori Mountains in equatorial Africa', *Ecological Monographs* 37, 25–52.

LLOYD, J. W. (1973) 'Climatic variations in north central Chile from 1866 to 1971', *Journal of Hydrology* 19, 53–70.

LONGWELL, C. R. (1960) 'Interpretation of the levelling data', *United States Geological Survey Professional Paper* 295, 33–8.

LYALL, I. T. (1970) 'Recent trends in spring weather', *Weather* 25, 163–5.

MABBUTT, J. A. (1971) 'The Australian arid zone as a prehistoric environment', in Mulvaney, D. J. and Golson, J. (eds.), *Aboriginal man and environment in Australia* (ANU, Canberra), 66–79.

MALDE, H. E. (1964) 'Environment and man in arid America', *Science* 145, 123–9.

MANLEY, G. (1953) 'The mean temperature of central England 1698–1952', *Quarterly Journal of The Royal Meteorological Society* 79, 242–61.

MANLEY, G. (1964) 'The evolution of the climatic environment', in Watson, W. and Sissons, J. B. (eds.), *The British Isles: a systematic geography* (Nelson, London).

MANLEY, G. (1966) 'Problems of the climatic optimum: the contribution of glaciology', in *Royal Meteorological Society Symposium on world climate 8000 to 0 B.C.* 34–9.

MANLEY, G. (1971) 'Interpreting the meteorology of the Late and Post-Glacial', *Palaeogeography Palaeoclimatology Palaeo-ecology* 10, 163–75.

MANLEY, G. (1974) 'Central England temperatures: monthly means 1659–1973', *Quarterly Journal of the Royal Meteorological Society* 100, 389–405.

MARKGRAF, V. (1974) 'Paleoclimatic evidence derived from timberline fluctuations', *Colloques Internationaux du CNRS* 219, 67–76.

MARTIN, P. S. (1963) 'Early man in Arizona: the pollen evidence', *American Antiquity*, 29, 67–73.

MARTIN, P. S. (1966) 'Africa and Pleistocene overkill', *Nature* 212, 339–42.

MARTIN, P. S. (1967) 'Prehistoric overkill', in Martin, P. S. and Wright, H. E. (eds.), *Pleistocene extinctions* (Yale University Press, New Haven), 75–120.

MASON, B. J. (1976) 'The nature and prediction of climatic changes', *Endeavour* 35, 51–7.

MASURIER LE, W. E. (1972) 'Volcanic record of Antarctic glacial history: implications with regard to Cenozoic Sea-Levels', in Price, R. J. and Sugden, D. E. (eds.), *Polar geomorphology (Institute of British Geographers Special Publication* No. 4, 59–74).

MATHEWS, W. H. and CURTIS, G. H. (1966) 'Date of the Plio-Pleistocene boundary in New Zealand', *Nature* 212, 979–80.

MAYR, F. (1964) 'Untersuchungen über Ausmass und Golgen der Klima- und Gletscherwankungen seit dem Beginn der post-glazialen Wärmezeit', *Zeitschrift für Geomorphologie* 8, 257–85.

MCBURNEY, C. B. M. and HEY, R. W. (1955) *Prehistory and Pleistocene geology in Cyrenaican Libya* (Cambridge U.P., Cambridge).

MCCREA, W. H. (1975) 'Ice ages and the galaxy', *Nature* 255, 607–9.

MCDOUGALL, I. and STIPP, J. J. (1968) 'Isotopic dating evidence for the age of climatic deterioration and the Pliocene-Pleistocene boundary', *Nature* 219, 51–3.

MCDOUGALL, I. and WENSINK, H. (1966) 'Paleomagnetism and geochronology of the Pliocene-Pleistocene lavas in Iceland', *Earth and Planetary Science Letters* 1, 232–6.

MCINTYRE, A. and RUDDIMAN, W. F., and JANTZEN, R. (1972) 'Southward penetrations of the North Atlantic Polar Front: faunal and floral Evidence of large-scale surface water mass movements over the last 225 000 years', *Deep-Sea Research* 19, 61–77.

MCLURE, H. A. (1976) 'Radiocarbon chronology of Late Quaternary lakes in the Arabian desert', *Nature* 263, 755–6.

MCMASTER, L. E. and GARRISON, R. L. (1966) 'Morphology and sediments of the continental shelf of southern New England', *Marine Geology* 4, 273–89.

MEADOWS, A. J. (1975) 'A hundred years of controversy over sunspots and weather', *Nature* 256, 95–7.

MEGGERS, B. J. (1975) 'Application of the biological model of diversification to cultural distributions in tropical lowland South America', *Biotropica* 7, 141–61.

MEIER, M. F. (1965) 'Glaciers and climate', in Wright H. E. and Frey, D. G. (eds.), *The Quaternary of the U.S.A.*, 795–805.

MENGEL, R. M. (1970) 'The North American central plains as an isolating agent in bird speciation', in Dort, W. and Jones, J. K. (eds.), *Pleistocene and Recent environments of the central Great Plains* (Kansas U.P.), 279–340.

MERCER, J. H. (1969) 'The Allerød oscillation: a European climatic anomaly', *Arctic and Alpine Research* 1, 227–34.

MERCER, J. H. (1972) 'The lower boundary of the Holocene', *Quaternary Research* 2, 15–24.

MESOLLELA, K. J., MATTHEWS, R. K., BROECKER, W. S., and THURBER, D. L. (1969) 'The astronomical theory of climatic change: Barbados data', *Journal of Geology* 77, 250–74.

MICHEL, P. (1968) 'Genèse et Évolution de la Vallée du Sénégal de Bakel à l'Embouchure (Afrique Occidentale)', *Zeitschrift für Geomorphologie* 12, 318–49.

MICKLIN, P. P. (1972) 'Dimensions of the Caspian Sea problem', *Soviet Geography* 13, 589–603.

MILLER, C. D. (1969) 'Chronology of Neo-glacial moraines in the Dome Peak area, North Cascade Range, Washington', *Arctic and Alpine Research* 1, 49–66.

MILLIMAN, J. D. and EMERY, K. O. (1968) 'Sea-Levels during the past 35 000 years', *Science* 162, 1121–3.

MILTON, D. (1974) 'Some observations of global trends in tropical cyclone frequencies', *Weather* 29, 267–70.

MITCHELL, G. F. (1972) 'Soil deterioration associated with prehistoric agriculture in Ireland', *20th International Geological Congress* Symposium 1, 59–68.

MITCHELL, J. M. (1963) 'On the world-wide pattern of secular temperature change', *Arid Zone Research* 20, 161–81.

MITCHELL, J. M. (1965) 'Theoretical paleoclimatology', in Wright, H. E. and Frey, D. G. (eds.), *The Quaternary of the U.S.A.*, 881–901.

MITCHELL, J. M. (ed.) (1968) 'Causes of climatic change', *Meteorological Monographs* 8.

MONTFORD, H. M. (1970) 'The terrestrial environment during Upper Cretaceous and Tertiary Times', *Proceedings Geologists' Association* 81, 181–204.

MOORE, P. D. (1975) 'Origin of blanket mires', *Nature* 256, 267–9.

MOREAU, R. E. (1963) 'Vicissitudes of the African biomes in the Late Pleistocene', *Proceedings Zoological Society of London* 141, 395–421.

MÖRNER, N. A. (1969) 'The Late Quaternary history of the Kattegatt Sea and the Swedish west coast', *Sveriges Geologiska Undersökning* Series C, NR.640, Arsbok 63, NR.3.

MÖRNER, N. A. (1971) 'Eustatic and climatic oscillations', *Arctic and Alpine Research* 3, 167–71.

MÖRNER, N. A. (1971) 'The Plum Point Interstadial: age, climate and subdivision', *Canadian Journal of Earth Science* 8, 1423–31.

MÖRNER, N. A. (1972) 'Time-scale and ice accumulation during the last 125 000 years as indicated by the Greenland 0^{18} curve', *Geological Magazine* 109, 17–24.

MÖRNER, N. A. (1976) 'Eustasy and geoid changes', *Journal of Geology* 84, 123–51.

MORRISON, H. E. S. (1968) 'Pleistocene vegetation and climate in Uganda', *Journal of Ecology* 56, 363–84.

NEWELL, N. D. and BLOOM, A. L. (1970) 'The reef flat and "two-meter eustatic terrace" of some Pacific atolls', *Bulletin, Geological Society of America* 81, 1881–93.

NEWMAN, W. S., FAIRBRIDGE, R. W., and MARCH, S. (1971) 'Marginal subsidence of glaciated areas: United States, Baltic and North Seas', *Quaternaria* 14, 39–40.

NEWELL, N. D. (1961) 'Recent terraces of tropical limestone shores', *Zeitschrift für Geomorphologie Supplementband* 3, 87–106.

NICHOLS, H. (1967) 'Central Canadian palynology and its relevance to North-western Europe in the late Quaternary period', *Review of Palaeobotany and Palynology* 2, 231.

OLAUSSON, E. and OLSSON, I. V. (1969) 'Varve stratigraphy in a core from the Gulf of Aden', *Palaeogeography Palaeoclimatology Palaeo-ecology* 6, 87–103.

OPDYKE, N. D., GLASS, B., HAYS, J. D., and FOSTER, J. (1966) 'Paleomagnetic study of Antarctic deep-sea cores', *Science* 154, 349–57.

OPIK, E. J. (1958) 'Climate and the changing sun', *Scientific American* 198, 85–92.

OSBORNE, P. J. (1974) 'An insect of early Flandrian Age from Lea Marston, Warwickshire, and its bearing on the contemporary climate and ecology', *Quaternary Research* 4, 471–86.

OSMOND, J. K., CARPENTER, J. R., and WINDOM, H. L. (1965) 'Th^{230}/U^{234} Age of the Pleistocene corals and oolites of Florida', *Journal of Geophysics Research* 70, 1843–7.

OTTERMAN, J. (1974) 'Baring high albedo soils by overgrazing: a hypothesized desertification mechanism', *Science* 186, 531–3.

PARMENTER, C. and FOLGER, D. W. (1974) 'Eolian biogenic detritus in deep-sea sediments: a possible index of equatorial ice-age aridity', *Science* 185, 695–8.

PARKIN, D. W. and SHACKLETON, N. J. (1973) 'Trade winds and temperature correlations down a deep-sea core off the Saharan coast', *Nature* 245, 455–7.

PARRY, M. L. (1975) 'Secular climatic change and marginal agriculture', *Transactions Institute of British Geographers* 64, 1–13.

PATZELT, G. (1974) 'Holocene variations of glaciers in the Alps', *Colloques, Internationaux du CNRS* 219, 51–9.

PEARSALL, W. H. (1964) 'After the ice retreated', *New Scientist* 383, 757–9.

PENCK, A. and BRÜCKNER, E. (1909) *Die Alpen in Eiszeitalten* (Tauchnitz, Leipzig).

PENNINGTON, W. (1969) *The history of British vegetation* (English University Press, London).

PHILLIPS, C. S. (ed.) (1970) *The Fenland in Roman times* (Royal Geographical Society, London).

PILBEAM, D. R. (1975) 'Middle Pleistocene hominids', in Butzer, K. W. and Isaac, G. L. (eds.), *After the Australopithecines* (Mouton, The Hague), 809–56.

PLAFKER, G. (1965) 'Tectonic deformation associated with the 1964 Alaska earthquake', *Science* 148, 1675–87.

PLASS, G. N. (1956) 'The Carbon Dioxide theory of climatic change', *Tellus* 8, 140–54.

POTTER, G. L., ELLSAESSER, H. W., MACCRACKEN, M. C., and LUTHER, F. M. (1975) 'Possible climatic impact of tropical deforestation', *Nature* 258, 697–8.

PRANCE, G. T. (1973) 'Phytogeographic support for the theory of Pleistocene forest refuges in the Amazon basin, based on evidence from distribution patterns in Caryocaraceae, Chrysobalanaceae, Dichapetalaceae and Lecythidaceae', *Acta Amazonica* 3, 5–28.

PRICE, W. A. (1958) 'Sedimentology and Quaternary geomorphology of South Texas', *Transactions Gulf Coast Association of Geological Societies* 8, 41–75.

RAIKES, R. (1972) *Water weather and prehistory* (Baker, London), pp. 208ff.

RAPP, A. (1974) 'A Review of desertization in Africa—water vegetation, and man', *Secretariat for International Ecology* (Stockholm, Report No. 1.).

RASOOL, S. I. and SCHNEIDER, S. H. (1971) 'Atmospheric Carbon Dioxide and aerosols: affects of large increases on global climate', *Science* 173, 138–41.

RECK, R. A. (1975) 'Aerosols and solar temperature changes', *Science* 188, 728–30.

REED, C. A. (1970) 'Extinction of mammalian megafauna in the Old World late Quaternary', *Bioscience* 20, 284–8.

REEVES, C. C. (1966) 'Pluvial lake basins of west Texas', *Journal of Geology* 74, 269–91.

Report of the Study of Critical Environmental Problems (1970) *Man's impact on the global environment* (M.I.T. Press, Cambridge), pp. 319ff.

RICHMOND, G. M. (1970) 'Comparison of the Quaternary stratigraphy of the Alps and Rocky Mountains', *Quaternary Research* 1, 3–28.

RIEHL, H. (1956) 'Sea surface temperature anomalies and hurricanes', *Bulletin American Meteorological Society* 37, 413–17.

RIPLEY, E. A. (1976) 'Drought in the Sahara: insufficient geophysical feedback?', *Science* 191, 100.

RODDA, J. C. (1970) 'Rainfall excesses in the United Kingdom', *Transactions, Institute of British Geographers* 49, 49–60.

ROGNON, P. (ed.) (1976) 'Oscillations climatiques au Sahara depuis 40 000 ans', *Revue de Géographie physique et de géologie dynamique*, 18, 147–282.

RONA, E. and EMILIANI, C. (1969) 'Absolute dating of Caribbean cores P6304–8 and P6304–9', *Science* 163, 66–8.

RONAI, A. (1965) 'Neo-tectonic subsidence in the Hungarian basin', *Geological Society of American special Paper* 84, 219–32.

ROSENAN, N. (1963) 'Climatic fluctuations in the Middle East during the period of instrumental record', *Arid Zone Research* 20, 67–73.

RUSSELL, R. J. (1967) 'Aspects of coastal morphology', *Geografiska Annaler* 49, 299–309.

RUSSELL, F. S., SOUTHWARD, A. J., BOALCH, G. T., and BUTLER, E. I. (1971) 'Changes in biological conditions in the English Channel off Plymouth during the last half century', *Nature* 234, 468–70.

SAARNTHEIN, M. (1972) 'Sediments and history of the Post-glacial transgression in the Persian Gulf and north-west Gulf of Oman', *Marine Geology* 12, 245–66.

SALINGER, M. J. (1976) 'New Zealand temperatures since 1300 A.D.', *Nature* 260, 310–11.

SALINGER, M. J. and GUNN, J. M. (1975) 'Recent climatic warming around New Zealand', *Nature* 256, 396–8.

SAUER, C. O. (1948) 'Environment and culture during the last deglaciation', *Proceedings American Philosophical Society* 92, 65–77.

SAUER, C. O. (1968) *Northern mists* (University of California Press), pp. 204ff.

SAWYER, J. S. (1971) 'Possible effects of human activity on world climate', *Weather* 26, 251–62.

SCHELL, I. I. (1961) 'Recent evidence about the nature of climatic changes and its implications', *Annals, New York Academy of Science*, 95, 251–70.

SCHELL, I. I. (1962) 'On the iceberg severity off Newfoundland and its prediction', *Journal of Glaciology* 4, 161–72.

SCHELL, I. I. (1974) 'On the lag in the response of the ocean during a climatic change', *Climatic Research Unit Research Publication* 2, 85–93.

SCHOFIELD, J. C. (1960) 'Sea-level fluctuations during the past four thousand years', *Nature* 185, 836.

SCHOLL, D. W. (1964) 'Recent sedimentary record in mangrove swamps and rise in sea-level over the western coast of Florida', *Marine Geology* 1, 344–66.

SCHUMM, S. A. (1963) 'The disparity between present rates of denudation and orogeny', *United States Geological Survey Professional Paper* 454-H, pp. 13ff.

SEDDON, B. (1971) *Introduction to biogeography* (Duckworth, London). '

SEGOTA, T. (1966) 'Quaternary temperature changes in central Europe', *Erdkunde* 20, 110–18.

SEGOTA, T. (1967) 'Palaeotemperature changes in the Upper and Middle Pleistocene', *Eiszeitalte und Gegenwart* 18, 127–41.

SELBY, M. J. (1973) 'Antarctica: the key to the Ice Age', *New Zealand Geographer* 29, 134–50.

SEREBRYANNY, L. R. (1969) 'L'Apport de la Radiochronométrie e l'étude de l'histoire tardi-Quaternaire des Régions de Glaciation Ancienne de la Plaine Russe', *Revue de Géographie Physique et Géologie Dynamique* 11, 293–302.

SHACKLETON, N. J. (1967) 'Oxygen isotopic analyses and Pleistocene temperatures reassessed', *Nature* 215, 15–17.

SHACKLETON, N. J. (1975) 'The stratigraphic record of deep-sea cores and its implications for the assessment of glacials, interglacials, stadials and interstadials in the Mid-Pleistocene', in Butzer, K. W. and Isaac, G. L. (eds.), *After the Australopithecines* (Mouton, The Hague), 1–24.

SHANNAN, L., EVENARI, M., and TADMOR, N. H. (1967) 'Rainfall patterns in the central Negev Desert', *Israeli Exploration Journal* 17, 163–84.

SHEPARD, F. P. (1963) 'Thirty-five thousand years of sea-level', in Clements, T. (ed.), *Essays in marine geology in honor of K. O. Emery* (California U.P., Los Angeles).

SHOTTON, F. W. (1966) 'Problems and contributions of methods of absolute dating within the Pleistocene Period', *Quarterly Journal of the Geological Society* 122, 357–83.

SINGH, G. (1971) 'The Indus valley culture seen in the context of Post-Glacial climate and ecological Studies in north-west India', *Archaeology and Anthropology in Oceania* 6, 177–89.

SLAUGHTER, B. H. (1967) 'Animal ranges as a clue to Late Pleistocene extinction', in Martin, S. S. and Wright, H. E. (eds.), *Pleistocene Extinctions*, 155–67.

SLAYMAKER, H. O., HOWE, G. M., and HARDING, D. M. (1967) 'Some aspects of the flood hydrology of the upper catchments of the Severn and Wye', *Transactions, Institute of British Geographers* 41, 33–58.

SMALLEY, I. J. and VITA-FINZI, C. (1968) 'The formation of fine particles in sandy deserts and the nature of "desert" loess', *Journal of Sedimentary Petrology* 38, 766–74.

SMITH, A. G. (1970) 'The influence of Mesolithic and Neolithic man on British vegetation: a discussion', in Walker, D. and West, R. G. (eds.), *The vegetational history of the British Isles* (Cambridge University Press, Cambridge), 81–96.

SMITH, C. G. (1967) 'Winters at Oxford since 1815', *Oxford Magazine*, March, pp. 4ff.

SMITH, H. T. U. (1965) 'Dune morphology and chronology in central and western Nebraska', *Journal of Geology* 73, 557–78.

SMITH, P. J. (1973) *Topics in geophysics* (Open University Press, Bletchley).

SNYDER, C. T. and LANGBEIN, W. B. (1962) 'The Pleistocene lake in Spring Valley, Nevada and its climatic implications', *Journal of Geophysical Research* 67, 2385–94.

SOLECKI, R. (1963) 'Prehistory in Shanidar Valley, northern Iraq', *Science* 139, 179–83.

SPARKS, B. W. and WEST, R. G. (1972) *The Ice Age in Britain* (Methuen, London), pp. 302ff.

STEARNS, C. E. and THURBER, D. L. (1967) 'Th^{230}/U^{234} Dates of Late Pleistocene marine fossils from the Mediterranean and Moroccan littorals', *Progress in Oceanography* 4, 293–305.

STEENSBERG, A. (1951) 'Archaeological dating of the climatic change in north Europe about A.D. 1300', *Nature* 168, 672–4.

STEPHENSON, R. L. (1965) 'Quaternary human evolution of the Plains', in Wright, H. E. and Frey, D. G. (eds.), *The Quaternary of the U.S.A.*, 685–707.

STODDART, D. R. (1969) 'Sea-level change and the origin of sand cays; radiometric evidence', *Journal of the Marine Biological Association of India*, 11, 44–58.

STODDART, D. R. (1971) 'Environment and history in Indian Ocean reef morphology', *Symposium, Zoological Society of London*, 28, 3–38.

STODDART, D. R. (1973) 'Coral reefs: the last two million years', *Geography* 58, 313–23.

STREET, F. A. and GROVE, A. T. (1976) 'Environmental and climatic implications of Late Quaternary lake level fluctuations in Africa', *Nature* 261, 385–90.

STRIDE, A. H. (1959) 'On the origin of the Dogger Bank in the North Sea', *Geological Magazine* 96, 33–44.

STRONGFELLOW, M. F. (1974) 'Lightning incidence in Britain and the solar cycle', *Nature* 249, 332–3.

TALLIS, J. H. (1975) 'Tree remains in southern Pennine peats', *Nature* 256, 483–4.

TANNER, W. F. (1968) 'Tertiary sea-level Symposium—introduction', *Palaeogeography Palaeoclimatology and Palaeo-ecology* 5, 7–14.

TARLING, D. H. and TARLING, M. H. (1971) *Continental drift* (Bell, London).

TAUBER, H. (1970) 'The Scandinavian varve chronology and C14 dating', in Olsson, I. U. (ed.), *Radiocarbon variation and absolute chronology* (Wiley, New York), 173–96.

THEAKSTONE, W. H. (1965) 'Recent changes in the Glaciers of Svartisen', *Journal of Glaciology* 5, 411–31.

THOM, B. G. (1973) 'The dilemma of high interstadial sea-levels during the Last Glaciation', *Progress in Geography* 5, 167–246.

THOM, B. G., HAILS, J. R., and MARTIN, A. R. H. (1969) 'Radiocarbon evidence against Post-Glacial higher sea-levels in eastern Australia', *Marine Geology* 7, 161–8.

THOMSON, J. and WALTON, A. (1972) 'Redetermination of chronology of Aldabra atoll by $^{230}Th/^{234}U$ Dating', *Nature* 240, 145.

THORARINSSON, S. (1940) 'Recent glacier shrinkage and eustatic changes of sea-level', *Geografiska Annaler* 22, 131–59.

THURBER, D. L. *et al.* (1965) 'Uranium-Series ages of Pacific atoll coral', *Science* 149, 55–8.

TJIA, H. D. (1970) 'Rates of diastrophic movement during the Quaternary in Indonesia', *Geologie en Mijnbouw* 49, 335–8.

TRICART, J. (1974) 'Existence de Périodes Seches au Quaternaire en Amazonie et dans les Régions Voisines', *Revue de Géomorphologie Dynamique* 23, 145–58.

TROELS-SMITH, J. (1956) 'Neolithic period in Switzerland and Denmark', *Science* 124, 876–9.

TUCKER, G. B. (1975) 'Climate: is Australia's changing?', *Search* 6, 323–8.

TURNER, C. and WEST, R. G. (1968) 'The subdivision and zonation of interglacial periods', *Eiszeitalter und Gegenwart* 19, 93–101.

TURNER, J. (1962) 'The *Tilia* decline: an anthropogenic interpretation', *New Phytologist* 61, 328–41.

TWIDALE, C. R. (1971) *Structural Landforms* (MIT Press, Cambridge, Mass.), pp. 247ff.

TWIDALE, C. R. (1972) 'Landform development in the Lake Eyre region, Australia', *Geographical Review* 62, 40–70.

United States Committee for the Global Atmospheric Research Program (1975) *Understanding climatic change, a program for action* (National Research Council, Washington D.C.).

UTTERSTRÖM, G. (1955) 'Climatic fluctuations and population problems in early modern history', *Scandinavian History Review* 3, 1–47.

VALENTINE, J. W. and VEEH, H. H. (1969) 'Radiometric ages of Pleistocene terraces from San Nicolas Island, California', *Bulletin Geological Society of America* 80, 1415–18.

VAN VEEN J. (1954) 'Tide-gauges, subsidence-gauges and flood-stones in the Netherlands', *Geologie en Mijnbouw* 33, 214–19.

VEEH, H. H. and CHAPPELL, J. (1970) 'Astronomical theory of climatic change: support from New Guinea', *Science* 167, 862–5.

VEEH, H. H. and GIEGENGACK, R. (1970) 'Uranium-Series ages of corals from the Red Sea', *Nature* 226. 155–6.

VEEH, H. H. and VALENTINE, J. W. (1967) 'Radiometric ages of Pleistocene fossils from Cayucos, California', *Bulletin Geological Society of America*, 78, 547–9.

VEEH, H. and VEEVERS, J. J. (1970) 'Sea-level at − 175 M off the Great Barrier Reef, 13 600 to 17 000 years ago', *Nature* 226, 526–7.

VEENSTRA, H. J. (1970) 'Quaternary North Sea coasts', *Quaternaria* 12, 169–79.

VERSTAPPEN, H.Th. (1970) 'Aeolian geomorphology of the Thar Desert and palaeoclimates', *Zeitschrift für Geomorphologie, Supplementband* 10, 104–20.

VERSTAPPEN, H. TH. (1975) 'On palaeo climates and landform development in Malesia', in Bartstra G.-J., and Casparie, W. A. (eds.), *Modern Quaternary research in Southeast Asia* (Balkema, Rotterdam), 3–35.

VIBE, C. (1967) 'Arctic animals in relation to climatic fluctuations', *Meddelelser Groenland* 170, pp. 227ff.

VITA-FINZI, C. (1969) *The Mediterranean valleys* (Cambridge University Press, Cambridge), pp. 140ff.

VITA-FINZI, C. (1973) *Recent earth history* (Macmillan, London), pp. 130ff.

VIVIAN, R. (1975) *Les Glaciers des Alpes Occidentales* (Imprimerie Allier, Grenoble).

VUILLEUMIER, B. S. (1971) 'Pleistocene changes in the fauna and flora of South America', *Science* 173, 771–80.

WALTON, K. (1966) 'Vertical movements of shorelines in highland Britain: an introduction', *Transactions, Institute of British Geographers* 39, 1–8.

WEARE, B. C., TEMKIN, R. L., and SNELL, F. M. (1974) 'Aerosols and climate: some further considerations', *Science* 186, 827–8.

WEGMANN, E. (1969) 'Changing ideas about moving shorelines', in Scheer, C. J. (ed.), *Toward a history of geology* (M.I.T. Press, Cambridge), 386–414.

WENDORF, F., Schild R., SAID R., HAYNES, C. V., GAUTIER, A., and KOBUSIEWICZ, P. (1976), 'The prehistory of the Egyptian Sahara', *Science* 193, 103–16.

WEST, R. G. (1972) *Pleistocene geology and biology* (Longman, London).

WESTERN, D. and PRAET, C. V. (1973) 'Cyclical changes in the habitat and climate of an East African ecosystem', *Nature* 241, 104–6.

WEXLER, H. (1961) 'Additional comments on the warming trend at Little America, Antarctica', *Weather* 16, 56.

WEYL, P. K. (1968) 'The role of the oceans in climatic change: a theory of the ice ages', *Meteorological Monographs* 8, 37–62.

WHITTOW, J. B. (1973) 'Shoreline evolution and the eastern coast of the Irish Sea', *Quaternaria* 12, 185–96.

WHITTOW, J. B., SHEPHERD, A., GOLDTHORPE, J. E., and TEMPLE, P. H. (1963) 'Observations on the glaciers of the Ruwenzori', *Journal of Glaciology* 4, 581–616.

WHYTE, R. O. (1963) 'The significance of climatic change for natural vegetation and agriculture', *Arid Zone Research* 20, 381–93.

WICKENS, G. E. (1975) 'Changes in the climate and vegetation of the Sudan since 20 000 B.P.', *Boissiera* 24, 43–65.

WILLIAMS, G. E. and POLACH, H. A. (1971) 'Radiocarbon dating of arid zone calcareous paleosols', *Bulletin Geological Society of America* 82, 3069–86.

WILLIAMS, J. (1975) 'The influence of snowcover on the atmospheric circulation and its role in climatic change: an analysis based on results from the near global circulation model', *Journal of Applied Meteorology* 14, 137–52.

WILLIAMS, R. B. G. (1975) 'The British climate during the Last Glaciation; an interpretation based on periglacial phenomena', in Wright, A. E. and Moseley, F. (eds.), *Ice ages: ancient and modern* (Seel House, Liverpool), 95–127.

WILLIS, E. H. (1961) 'Marine transgression sequences in the English Fenlands', *Annals New York Academy of Sciences* 95, 368–76.

WILSON, A. T. (1964) 'Origin of ice ages: an ice shelf theory for Pleistocene glaciation', *Nature* 201, 147–9.

WINSTANLEY, D. (1973) 'Rainfall patterns and general atmospheric circulation', *Nature* 245, 190–4.

WOLLIN, G., ERICSON, D. B., and RYAN, W. B. F. (1971) 'Variations in magnetic intensity and climatic changes', *Nature* 232, 549–51.

WOLLIN, G., ERICSON, D. B., and WOLLIN, J. (1974) 'Geomagnetic variations and climatic change, 2 000 000 B.C.–1970 A.D.', *Colloques Internationaux du CNRS* 219, 273–86.

WOLLIN, G., KUKLA, G. J. ERICSON, D. B., RYAN, W. B. F., and WOLLIN, J. (1973) 'Magnetic intensity and climatic changes 1925–1970', *Nature* 242, 34–6.

WOOD, C. A. and LOVETT, R. R. (1974) 'Rainfall, drought and the solar cycle', *Nature* 252, 594–6.

WOODBURY, R. B. (1961) 'Climatic changes and prehistoric agriculture in the South-western United States', *Annals New York Academy of Sciences* 95, 705–9.

WRIGHT, H. E. (1968) *Climatic change in the eastern Mediterranean region: the natural environment of early food production in the mountains north of Mesopotamia.* Final report, University of Minnesota, Contract NONR—710 (33) Task No. 380–129.

WRIGHT, H. E. and FREY, D. G. (1965) (eds.) *The Quaternary of the United States* (Princeton U.P.).

WRIGHT, W. B. (1937) *The Quaternary ice age* (Macmillan, London).

WYRWOLL, K.-H. and MILTON, D. (1976) 'Widespread late Quaternary aridity in western Australia', *Nature* 264, 429–30.

ZAGWIJN, W. H. (1974) 'The Pliocene-Pleistocene boundary in western and southern Europe', *Boreas* 3, 75–97.

ZAGWIJN, W. H. (1975) 'Variations in climate as shown by pollen-analysis especially in the Lower Pleistocene of Europe', in Wright, A. E. and Moseley, F. (eds.), *Ice ages: ancient and modern* (Seel House, Liverpool), 137–52.

ZEUNER, F. E. (1959) *The Pleistocene period: its climate, chronology and faunal successions* 2nd edn. (Hutchinson, London).

VAN ZINDEREN BAKKER, E. M. (1962) 'A Late Glacial and Post-Glacial correlation between East Africa and Europe', *Nature* 194, 201.

ZUBAKOV, V. A. (1969) 'La Chronologie des Variations Climatiques au Cours du Pleistocene en Sibérie Occidentale', *Revue de Géographie Physique et Géologie Dynamique* 11, 315–24.

Index